Wild American Ginseng

Wild American Ginseng

LESSONS FOR CONSERVATION IN THE AGE OF HUMANS

James McGraw

The University of Georgia Press
Athens

Published by the University of Georgia Press
Athens, Georgia 30602
www.ugapress.org
© 2023 by James McGraw
All rights reserved
Designed by Kaelin Chappell Broaddus
Set in 10.5/13.5 Garamond Premier Pro Regular
by Kaelin Chappell Broaddus
Printed and bound by Sheridan Books, Inc.
The paper in this book meets the guidelines for permanence
and durability of the Committee on Production Guidelines
for Book Longevity of the Council on Library Resources.

Most University of Georgia Press titles are
available from popular e-book vendors.

Printed in the United States of America
27 26 25 24 23 P 5 4 3 2 1

Library of Congress Cataloging-in-Publication Data

Names: McGraw, James B., 1956– author.
Title: Wild American ginseng : lessons for conservation
in the age of humans / James McGraw
Other titles: Lessons for conservation in the age of humans
Description: Athens, Georgia : The University of Georgia Press,
[2023] | Includes bibliographical references and index.
Identifiers: LCCN 2022017995 | ISBN 9780820362670
(hardback) | ISBN 9780820361987 (paperback)
| ISBN 9780820362687 (ebook)
Subjects: LCSH: Ginseng. | Plant conservation.
Classification: LCC SB295.G5 M34 2023 | DDC
639.9/9—dc23/eng/20220629
LC record available at https://lccn.loc.gov/2022017995

To my father, who modeled for his sons how to reason like scientists yet embrace the wonder to be found in music and unspoiled nature

CONTENTS

Wild American Ginseng

CHAPTER 1

The Panacea

We have not inherited the earth from our ancestors;
we have borrowed it from our children.—**ANONYMOUS**

Sweltering. That was the only word that could adequately describe that August day in western Kentucky. We waded through a waist-high stand of tick-filled grass and headed toward the woods. My graduate student Jen, normally unflappable in the face of stress, looked worried. At 102 degrees and humid, we were both dripping sweat within seconds of leaving the car. I was concerned about her. And I suspected she was wondering who her new advisor might be after the preceding one had keeled over from heatstroke. "Discretion is the better part of valor?" I queried. She nodded. We retreated to the rooms of our air-conditioned hotel and returned after dinner to complete our fieldwork, by which time the temperature would drop to a mere 92.

What madness could compel us to pursue ecological fieldwork under such conditions? American ginseng. We were six hundred miles from home on our "western circuit" to census natural populations of this unassuming little plant. But not just any populations. These were populations that my graduate students and I had already been censusing, and sweating over, twice per year, for fourteen years, and we were not about to let another spell of hot weather break the data stream. This particular population was especially fascinating. A couple of years earlier, a winter ice storm had destroyed much of the overstory tree canopy, leaving plants exposed to intense sun. Could they survive this newly created understory desert? If not, what did that mean for the future of the plant under a warming climate?

But let's back up. Why had ginseng captured our attention in the first place? Indeed, why should anyone care what happens to this very ordinary-looking, low-statured green herb that is widely scattered across much of eastern North America? For the answer to that, we need to go back in history a few years—

or perhaps a few thousand. By two thousand years ago, ginseng was firmly established in the herbal pharmacopeia of China. Its reputation only grew as it became scarcer, eventually being reserved only for emperors and other members of the royal family. But this was Asian ginseng, a different species altogether. Why did we study its close cousin, American ginseng, also being used two thousand years ago, by indigenous North Americans?

A clue to the significance of ginseng to Asian consumers is found in the genus name endowed by Linnaeus—*Panax*. This species was believed to be a panacea, a cure-all. Eastern medicinal practices developed long before the formal scientific method and also before detailed studies of human anatomy and physiology made Western medicine possible. The language of traditional herbal medicine seems strange to us now. Translating from the first compendium of traditional herbal knowledge, the *Shennong Bencao Jing*, the author Subhuti Dharmananda says this about ginseng in an article in *HerbalGram*: "Ginseng is sweet and a little cold. It mainly supplements the five viscera. It quiets the essence spirit, settles the ethereal and corporeal souls, checks fright palpitations, eliminates evil qi, brightens the eyes, opens the heart, and sharpens the wits. Protracted taking may make the body light and prolong life."

This is the powerful but mysterious language of mystics and shamans. While westerners may get an inkling of what powers ginseng possessed in the eyes of the ancient Chinese from these words, Dharmananda goes on to translate further. Ginseng was believed to tonify the spleen, calm irritation, serve as a nutritive aid, quiet the mind, reduce the frequency of disturbing dreams, strengthen the body, reduce heart palpitations, reduce anxiety and panic attacks, protect the body from disease, help the body recover from disease, keep the mind bright and alert, encourage happiness, stimulate courage and purpose, and prolong life. Still not always the language of science, but what a list! A panacea indeed.

But are two-thousand-year-old beliefs, rooted in shamanistic traditions, enough to make us care about ginseng today? Incredibly, the answer is yes. Versions of the old beliefs in ginseng's power persist in Asian cultures today. Traditional medicine is still practiced, either on its own or as a supplement to Western medicine. Therefore, demand for ginseng roots remains strong.

But why is *wild* ginseng coveted? Can't ginseng be cultivated like any other crop, leaving natural populations to prosper? Indeed, ginseng is cultivated, although it is a tricky crop to grow, requiring multiple years of care to reach harvestable size, expensive shade cloth structures to protect from sunburn, and intensive spraying with pesticides to prevent crop loss to which the plant is particularly prone when grown in dense monocultures. In addition, cultivated plants produce roots that are smooth and straight, like a carrot. In traditional Asian medicine, the gnarly, twisted roots of wild ginseng, especially if shaped remotely like a human body, are believed to be far more potent than cultivated roots. These wild roots command up to tenfold higher prices than cultivated roots. Next time you are in New York or San Francisco, stop in to a ginseng store in Chinatown and check out the prices of old, gnarly, wild ginseng roots. Your jaw will drop. Though certainly an extreme example, one sixty-five-year-old ginseng root shown mounted in a box in Tiananmen Square in Beijing (ca. 2013) had a list price of $588,000. Profit for a wild ginseng digger varies from year to year, but a batch of two hundred to three hundred dried roots weighing a pound will fetch between $200 and $800 at the dealer.

With Asian ginseng virtually extinct in the wild, the harvest of such coveted wild roots has been outsourced to North America, where ginseng persists in thousands of tiny populations spread across the eastern deciduous forest. Asian buyers from New York City and elsewhere purchase the roots from hundreds of dealers, who in turn buy their roots from thousands of harvesters who comb the woods each fall. After careful sorting, the Asian buyers eventually sell them across the Pacific Ocean to waiting consumers. As you will glean from chapter 2, because of its ongoing economic value, wild ginseng harvest has been a cultural touchstone in much of eastern North America since before the Revolutionary War. The ongoing market for wild ginseng played an important role in westward expansion. For tens of thousands, it served as an economic backstop for some of the most challenging regions of the country in which to make a viable living.

But what about ginseng's value in Western medicine? From aspirin (derived from willow) to morphine (poppy), plants have proven to be invaluable sources of active ingredients. The scientific approach to drug discovery involves isolating the bioactive compounds in plant tissues and carefully testing them in controlled studies with replication, including a placebo (control) group. In the United States, the Food and Drug Administration (FDA) approval process involves rigorous testing, often first using animal models, followed by extensive, replicated human trials. Has any compound from ginseng been isolated and tested, then adopted to treat specific diseases or conditions? The short answer is no. Does this mean ginseng is of no value to western medicine? Here the jury is far from reaching consensus.

Lots of careful studies, and some not so careful, have been published that represent the first tentative steps toward possible widespread adoption of ginseng-derived compounds as medicines. These studies have addressed possible uses of specific ginsenosides (the putative active ingredients in ginseng) to boost immune response, suppress cancer cell growth (prostate, breast), reduce diabetes symptoms, suppress appetite and induce weight loss, enhance memory, aid in recovery from brain injury, treat symptoms of depression, reduce pain sensitivity, prevent neural degeneration (Alzheimer's), increase physical stamina, reduce allergic reactions, and reduce inflammation, among many others. This list, not just by coincidence, has a strong overlap with the translated list from the *Shennong Bencao Jing*.

More than 150 naturally occurring ginsenosides have been isolated from ginseng plant parts. There may be many more. Which of these actually have effects? Are different ginsenosides effective for different purposes? In traditional medicine, ginseng is consumed not as isolated compounds but as whole tissue. Are other compounds in ginseng roots bioactive? We know that ginseng contains different forms of ginsenoside depending on where it grows—could there be forms of ginsenoside that have not even been discovered yet that are more effective than those tested already? Perhaps by Western standards ginseng is ineffective at treating half of the symptoms mentioned above. Maybe three-fourths. Successful drug development to address the remaining diseases and conditions would still relieve tremendous amounts of suffering. We have a huge amount left to learn about the medicinal properties of ginseng.

Even using standards of Western medicine, wild ginseng is already known to be a biochemical storehouse of potentially wonderful drugs to address some of the most pressing human health issues today. Ginsenosides most likely evolved as antifungal compounds in response to disease challenges experienced by natural populations over millions of years. The by-products of that evolutionary laboratory are the ginsenosides we examine today.

It might be useful to reframe the argument to illustrate the value of wild ginseng. Imagine that a single ginseng population (but we don't know which one!) contains a form of ginsenoside that is in fact a true cure for Alzheimer's disease (or breast cancer, or depression, or . . . you get the idea). Now, how much is the preservation of that population containing that one ginsenoside worth? Are we willing to risk losing it before we even know it exists? What price would we pay, as a society, to preserve that genotype?

So far, I have made the case that ginseng is economically and culturally important in eastern North America, highly prized in traditional Asian medicine, and potentially a source of drugs for common afflictions today. But what role does ginseng play in the forest community? Should it be preserved also because it is special in some way to other organisms that we care about? Unequivocally yes. First, ginseng foliage is consumed by a variety of insects but also by white-tailed deer, so it is part of the base of the food chain in the forest. Though its flowers are quite small, several pollinators visit and obtain nectar from them. Finally, birds and small mammals consume the fleshy red fruits. Research has even linked ginseng to one of our most beloved songbirds. Thus, while ginseng is indeed a small part of the plant biomass of a forest, it supports the animal community in the food web, adding richness to the whole ecosystem.

Twenty years ago, ongoing harvest led us to ask a simple question: "Is wild harvest of ginseng sustainable?" In other words, were wild harvesters practicing good stewardship of the resource? Or were harvested natural populations declining? The federal government wondered the same thing when it listed ginseng on an international treaty that required oversight of the harvest and international trade (CITES—more about this in chapter 2), giving a measure of legal protection to the plant by regulating it.

As my lab set out to establish a set of representative populations to census in the late 1990s and early 2000s, aspects of harvest—rates, effects, sustainability—were foremost in our thinking. But as we returned each year to these populations, it soon became obvious that harvest was not the only stressor on natural populations. We saw tops disappearing between our spring and fall visits, some clearly being nipped off by something, others disappearing for no evident reason. We saw plants turning yellow and dying back in midsummer. We saw evidence of hailstorms, late spring frost, ice storms, lightning strikes, flooding, and massive overstory defoliation. We saw aggressive invasive plant species overtopping them. We observed clear-cutting and selective harvest of trees that exposed plants to new stresses. And as we traveled from population to population, we noted landscapes permanently altered by strip-mining and housing developments. Over time, the larger picture was coming into focus, and from ginseng's vantage point it was not pretty.

Every wild plant or animal struggles for existence. Basic ecological principles tell us that overproduction of offspring combined with limited resources virtually guarantees intense competition. In addition, herbivores consume leaves. Frugivores consume fruits and seeds. Disease organisms infect plant tissues in their own struggle to survive. By carefully observing the forces acting on ginseng, following individuals over time from birth to death, the inevitable conclusion we reached was that *the struggle was being made much more difficult by the direct and indirect actions of our own species*. The more we learned, the more we began to see that ginseng was much more than America's most valuable wild-harvested medicinal plant. It was also a tangible, visible "canary in the coal mine" for the Anthropocene—the "era of humans" in which the effect of

FIGURE 1.1.
Ginseng-centric view of the "environment"

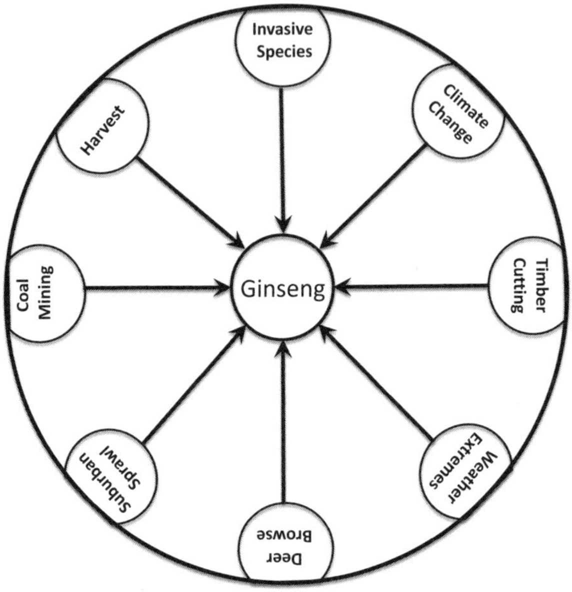

our species on the planet is so great, so all-encompassing, that it influences the natural and physical world of our planet in long-lasting ways. The long-term scientific studies of ginseng gradually poked holes in the blinders that had been preventing us from seeing this bigger picture. Taking a ginseng-centric point of view (fig. 1.1) opened our eyes to the extraordinary, human-influenced world of a rather ordinary plant.

The story of wild ginseng therefore evolved and became much more important than we originally thought. What was a straightforward question about harvest sustainability became a much broader, more significant, and more complex narrative about the relationship of our own species to wild species. This realization was only magnified as we saw parallels between ginseng and a host of species reported to be in decline not only by fellow botanists but also by scientists specializing in amphibians, bees, butterflies, and birds. A quarter-century ago, recognizing the cultural and economic significance of wild harvest, we could envision simply optimizing harvest regulations and all would be well. No more.

In the Anthropocene, evidence shows that *a new pact with nature's wild species is required to save biodiversity*. Wild American ginseng—this modest and beloved understory herb—beautifully illustrates why. This book weaves together the threads of scientific evidence, acquired through sweat, blood, and even a few tears, that support this conclusion. Along the way, I will dispel some long-held myths about ginseng, while raising questions that remain to be answered as we work to complete our tapestry of understanding. By the end of the book, I will sketch the outlines of possible futures for wild ginseng, seeking a path toward an optimistic outcome in which both humans and ginseng, indeed humans and wild species generally, can thrive.

CHAPTER 2

The Romantic Ideal

> Idealism and science continue to function in separate compartments; and yet "the happiness of man on earth" depends upon their combination.—**LEWIS MUMFORD**

Art and Ellie

August 16, 1975. "Be careful out there!" Ellie's mother yelled after her as she skipped up Papaw's driveway that morning. Today was going to be a special day in Ellie's summer vacation. She had been looking forward to it since June when Papaw had asked her if she thought she was big enough to go looking for wild ginseng with him.

"Can I?" she had said. "Really? Don't tease me if you don't mean it, Papaw!"

"I mean it, String Bean. If you think you're up for it. You have to be ready to get going at the crack of dawn. And you have to be able to hike all day up and down those steep hills with me. We might come up on a copperhead or a mean raccoon, and you better not fuss!"

"Ha. You know me, Papaw. I ain't afraid a no raccoon." She wasn't quite so sure about the copperhead, but she didn't want to admit it.

So the big day had arrived, and her mother had dropped her off at Papaw's. "You're late!" Papaw said, with a twinkle in his eye.

"I'm right on time," she exclaimed, feigning indignance. "Let's go!"

She hopped in his truck before he could get there. After taking his place behind the wheel, Papaw let loose the emergency brake and rolled the truck down the driveway with the engine off so he wouldn't wake Ellie's grandma. Then, with a flick of the key, they were off on their adventure.

Ten-year-old Ellie was the family tomboy, and, truth be told, Papaw was looking forward to this adventure at least as much as Ellie. So when he parked his truck on the side of the dirt road up Chestnut Holler, you would have been hard-pressed to find two happier human beings in all of McDowell County.

Now that he was retired from his job in the mines, Arthur Logan was freed to lavish his good nature upon his five grandchildren. He attended baseball games and school plays. But most of all he loved spending time in the outdoors with them. He taught them how to fish. He hoped to teach them how to hunt one day. But today was Ellie's day, and she was going to learn all about ginseng.

They quietly followed the dry creek bed upstream for a good mile before stopping for a rest on a boulder. "Where did all the water go, Papaw?"

"It's August, Ellie, and this stream only flows in the cool springtime when we've had lots of rain and the ground is wet. When it's hot out, the tree roots take up that water, and none of it reaches the stream, even when it does rain."

"Oh. I see. Those are some thirsty trees!" Ellie tipped over a rock, and a slippery salamander tried to squirm away. She grabbed it and took it to Papaw.

"Look at this! Do you know what kind he is, Papaw?"

"That's a slimy salamander, Ellie."

"I don't think he's slimy. He's just . . . slippery."

"Well, that's his official name: slimy salamander."

"I think they should have called him the spotted salamander. Look at those white spots!"

"You're right. That would be a better name, but common names don't always make sense."

She gently placed the salamander back on the ground, and then they hiked on, slowly climbing the gentle slope above the creek bed. Every once in a while, Papaw would show Ellie a new wildflower. "That's yellow root, Ellie. Some other day, we might dig some of that. But today we're after wild ginseng." "And that's black cohosh, Ellie. Check out that flower—it's almost as tall as you are!" "Oh, and this is pointer fern. When I see this one, I start to look around more carefully for ginseng, 'cause ofttimes I see them together." Ellie took it all in like a sponge. She would test herself so she would remember the names. This was fun! They kept hiking up the cove of hardwoods, eyes turned toward the ground.

"At last! Do you see it, String Bean?"

"See what?"

"Old man root! Ginseng."

"What? This little plant? What's so special about it?"

"That," Papaw replied with a note of pride in his voice, "is wild ginseng. Let's see if we can find some more. Look around and see if you can see others like it."

"Oh, here it is! Wait, and here's a big one over here, Papaw. And look! It has red berries! Now that *is* beautiful."

"You got it, Ellie. And that's one that we can dig. Those little ones, we'll leave for another year."

"But if we dig it, won't the plant die, Papaw? It's so perty. I don't want to kill it."

"Don't worry, Ellie. We're going to make sure she leaves plenty of baby ginseng around. That's why we dig only the big ones with bright red berries."

"Can I dig it, Papaw?"

"You can help, but let me show you how." He was worried she wouldn't be careful enough to get the whole root with all of its small root branches. So he demonstrated how to carefully clear away the leaf litter around the stalk, and prod the "neck" with her finger to see which direction it went. He explained how the neck would eventually lead to the root. Then he pulled an odd-looking tool out of his rucksack.

"What's that?" Ellie asked. She had never seen anything like it.

"That is my 'sang hoe. Watch. I loosen the soil all around the root, then pull away the clods with my fingers so I don't damage the root." He worked his way around the stalk, being careful not to nick the root.

After exposing the main root, Papaw found some deep tendrils were still stuck in the clay soil about six inches down, so he pulled a second tool out of his pack.

"What's that?"

"That's just a big ol' screwdriver, Ellie. I use this to loosen the deep soil and gently pull out the tiny roots so I keep our big momma root all in one piece." Then he held up the gnarled root like a trophy. "Look at that!"

"Oooh," Ellie said. "She's a beauty!"

"Yeah. This is a nice one. This root has two branches that look like legs. That's why it's called 'man root'!"

"But what about the babies, Papaw?"

"You don't miss a beat, do ya? I was so excited about the root, I almost forgot. Thanks, String Bean. OK, you can do this part. Take your pointer finger, and poke it in the ground about halfway to your knuckle, drop a berry in the hole, then fill it back in with fresh soil. Do the same thing with all of the berries. Spread them around so that when they come up, they will not be right on top of each other. That way, our big momma ginseng plant will have lots of babies long after she's gone."

Carefully following his instructions, she finished planting nine berries in the area around the hole where the momma plant was taken. "Wow, that was fun, Papaw!" "Let's find another plant to dig."

And so their day went, with Ellie eagerly seeking out ginseng plants with berries and learning to carefully step over the smaller plants. It was like a treasure hunt—Ellie had always loved those. She especially loved planting the bright red berries, imagining the day when a new plant would pop out of the ground right where she had buried the seed.

Ellie and her grandfather would remember that day for the rest of their lives. Ellie would remember the nature lessons she learned with a clarity she never achieved from schoolbooks. And from that day forward, Art Logan would always have a treasured memory of the day when he and his favorite plant were the center of his granddaughter's world. Art loved ginseng, because its gift *was* that day. Ellie loved ginseng for the same reason. How do you put a price on a plant that does that?

Cultural Touchstone

Arthur and Eleanor Logan are fictional characters. But in 1975 this was the picture folklorists and ethnobotanists painted of ginseng in Appalachia. Ginseng was a cure-all in more than one sense. The root connected self-educated rural naturalists to the rich forested landscape around them. But its powers were stronger than that. In the language of academic ethnobotanists, this unremarkable little herb forged indelible cross-generational bonds as eco-cultural knowledge was passed on. Even beyond these values, which are immense but impossible to quantify, were the economic benefits to these sustainable "wildcrafters" in the form of extra spending money collected in the season right before Christmas. Would Papaw dry the 150 roots he dug that year and sell them to a local dealer for $200, then buy all his grandchildren special gifts for Christmas? You bet he would. Ginseng was the gift that kept on giving.

Conservation is often equated with leaving nature alone. But conservation scientists were not heartless. Far from it. Despite concerns about ginseng's fate, they fervently wanted to preserve the ideal of a ginseng culture as much as ginseng itself, because of the very values that would be preserved along with the plant. They saw ginseng as a cultural keystone species that could further the larger goal of securing grassroots support for conservation of hundreds of lesser-known understory plants and the intact forests they required to thrive.

That same year, 1975, a seismic shift occurred in the wild ginseng world. Ginseng was listed under an international treaty to protect endangered animals and plants, the Convention on International Trade in Endangered Species of Wild Fauna and Flora (CITES). The listing recognized that wild ginseng was being harvested from the wild at high rates but more importantly that the sustainability of that harvest was questionable. At the very least, given the volume of trade over a long period of time, verification was needed that the harvest would not damage the wild resource.

But if ginseng harvest culture in 1975 indeed matched the story of Arthur Logan and his granddaughter, with careful stewardship of natural populations as a matter of course, how could ginseng possibly be endangered? Was it actually in decline? Indeed, how are we so sure ginseng populations were larger in the past, justifying the concern for its welfare? Answers to these questions are important because, with the CITES listing, the fate of ginseng became a national issue, subject to federal oversight. When this level of scrutiny is given to any natural resource, conflicts with states, localities, and especially individual rights are bound to arise. So the justification better be sound. To answer these questions, a historical look at ginseng is in order.

Origins of Trade

The story of ginseng in North America is sprinkled with a who's who of larger-than-life Americans, including John Jacob Astor, Thomas Jefferson, Daniel Boone, and John Muir. But we need to peer even farther back into history to truly understand ginseng's prior abundance on this continent. Ginseng was used by Native American peoples who were present for millennia prior to European colonization. From early written accounts, those in the forested parts of eastern North America were very familiar with the root, and many used it in a manner similar to their Asian counterparts in the temperate forest there. However, unlike in Asia, it was not particularly special, just one of dozens of plants in the apothecary of the time, certainly not the highly sought medicinal it later became when trade with Asia was established. Therefore, we can surmise that Native Americans had not depleted ginseng populations significantly in most of the range prior to colonization by Europeans.

Remarkably, overexploitation of wild ginseng populations can be traced to a single individual, a Jesuit missionary to the Mohawks in Quebec, which was part of New France, in 1715. Joseph-François Lafitau (fig. 2.1) had read a fellow missionary's writings about medicinal plants in China, and when he learned about the Chinese passion for ginseng, he set about looking for the plant in North America. Lafitau spent a good deal of time looking that first year, finally discovering a single plant with ripening berries in the fall. Within just five years, fur trappers and Native Americans fanned out across the forests of what is now southern Quebec and Ontario, digging the root and supplying the trade based

FIGURE 2.1. Joseph-Francois Lafitau (Portrait from Culture et Communications Quebec. Archives des Jesuites au Canada)

in Montreal. That certainly could not be done profitably today, even if it was legal. Indeed, a sharp decline in the availability of wild Canadian plants in the 1700s caused a shift in harvest to points south. The first regional boom-and-bust cycle in North America was over. Harvest and trade in wild ginseng are banned in Canada now, due to rarity that endures three centuries later. Only a handful of natural populations remain in each province and even fewer are considered viable. Clearly Canadian populations must have been much more numerous and larger in the early 1700s, assuming the trade records are correct about the volume of roots exported from Canada.

Westward Ho!

Records of trade kept by the young U.S. government in the late 1700s and early 1800s are also consistent with the notion that natural ginseng populations were much larger in the past than they are today. Through the first five U.S. administrations overseen by the Founders (Washington, Adams, Jefferson, Madison, and Monroe), ginseng was harvested at generally increasing rates (an average of 46,849 pounds per year during the Washington administration; 84,298 pounds annually during the Adams years; 202,199 pounds annually during the Jefferson years; 125,082 pounds annually during the Madison years; and 358,508 pounds annually during the Monroe presidency). Through much of the 1800s, export of ginseng topped 300,000 pounds per year. This suggests that the harvesters were spreading out into virgin ginseng territory over an expanding portion of the eastern deciduous forest.

This high-volume trade in ginseng played a significant role in American history and westward expansion after the Declaration of Independence. While the British had been shipping barrels of ginseng from its colony to China throughout the 1700s, after the Revolutionary War, the United States was looking to establish its own trade with Asian markets. Only seven years after Thomas Jefferson penned the Declaration of Independence, buyers were dispersed through Pennsylvania, Maryland, and Virginia to collect 60,000 pounds of wild root. In the spring of 1784, the *Empress of China* set sail for Canton, its hold filled with the treasured roots. It later returned with staggering profits for its investors. America's first millionaire, John Jacob Astor, best known for profiting from the fur trade, recognized this chance to expand his fortune, and by 1786 he was a major player in the ginseng trade with Asia. Thomas Jefferson, himself a tireless promoter of American natural resources, placed ginseng on his list of botanical treasures. Ginseng grew wild in the woods around Monticello, and no doubt Jefferson encountered it on his rambles in the woods of Virginia.

Frontier icon Daniel Boone had long been making money in the ginseng trade during the 1700s, but this venture was stymied during the Revolution. When he discovered trade had reopened in the new United States after the war, he redoubled his efforts to harvest wild ginseng and transport it to market. This led to the apocryphal story of Boone's loss of a huge quantity of ginseng roots when a barge capsized on the Ohio River. Whereafter, so the story goes, he shrugged at his bad luck and returned to the woods to gather more.

Booms and Busts

The westward expansion of the ginseng trade did not occur in one uniform wave. Instead, like a rising tide flowing into uneven tidepools, there were brief surges of intense harvest set against a background of ongoing exploitation elsewhere. Two well-documented examples of ginseng mania occurred in widely separated parts of the range. One illustrated a quick boom-and-bust scenario in a handful of years, while the second was drawn out over a few decades.

The rapid cycle was seen in the Big Woods region of south-central Minnesota in the far northwestern portion of the range. In 1858, Minnesota was experiencing a severe economic depression, and a large portion of its residents were deep in debt. Simultaneously, in a couple of small towns in the Big Woods, ginseng buyers showed up offering a high price for roots. That first year, only a few diggers headed out to look for the plant, but they scored big. By the following year, the word was out, and the ginseng rush was on as thousands of men, women, and children headed into the forest, shovels and picks in hand. They were rewarded handsomely, but within a few months the easily accessed populations were dug out. Nevertheless, the sudden influx of cash allowed thousands of Minnesotans to weather the worst of the economic crisis. Ginseng was hailed in local newspapers as a savior. Digging continued for a few more years, but finally the wild populations were reduced to levels that could not sustain the harvest. By 1865, the harvest had dropped by more than 50 percent, though national level exports had not declined.

The second boom-and-bust cycle was more drawn out and centered on the southern portion of the range, including western North Carolina, East Tennessee, western Virginia, eastern Kentucky, and southern West Virginia. By all accounts, then as today, this was the epicenter of wild ginseng country. The ginseng boom there was fueled by economic hardship as well. The region was devastated by the Civil War, and families needed to find alternative income sources wherever they could. John Muir stumbled on the world of the ginseng harvester on his famous thousand-mile hike from Indianapolis to Florida in the fall of 1867. Muir noted in his journal that a single county in East Tennessee produced $5,000 worth of ginseng that year (7,142 pounds at seventy cents a pound, $110 per pound in today's dollars). Thousands of diggers forayed into the woods each year, increasing the national harvest level to over 400,000 pounds per year into the 1880s. By the 1890s, the boom was over, as natural populations were depleted over a wide area. Assuming these plants were large relative to today's standards (perhaps one hundred roots per pound), that represents forty million plants being dug each year for a period of about thirty years. Over a billion plants! Clearly, ginseng was more abundant in the past than it is today.

To put the high harvest figures of the 1800s into perspective, in the twenty-six years from 1992 to 2017 (all post-CITES listing), the average annual export was only 77,369 pounds, about one-quarter of the average in the mid-1800s. That represents a mere 15.6 million roots per year (at 202 roots per pound, the U.S. Fish and Wildlife Service estimate). That is still a large number of roots, suggesting that probably tens of thousands of harvesters are taking to the woods annually. However, using harvest numbers as an indicator of the amount of ginseng in the woods makes a host of assumptions that may or may not be valid. First, it assumes that harvest pressure is the same. Harvest pressure is driven by demand from the Asian market, which is reflected in price. Of course, price needs to be adjusted for inflation. But even inflation-adjusted price is not enough to judge whether harvest pressure is the same. We know, for example, that when unemployment is high in Appalachia, ginseng harvest pressure goes up. The relative benefit to harvesters of going into the woods to look for ginseng increases dramatically when it becomes your only source of income, regardless of price. So the flow of ginseng in the medicinal plant trade depends on much more than "supply"; Asian economies determine demand, and local factors such as unemployment can determine whether the human capital exists to satisfy the demand. Economists have attempted to model this kind of complexity, and currently there are probably too many unknowns to be confident in their findings.

The Incredible Shrinking Plant

On top of all this, while we are certain that the high of 400,000 pounds of roots per year in the 1880s represents more plants than the 77,369 pounds per year of roots today, it is not five times as many. Old photographs show ginseng roots that were much larger, on average, than they are today (fig. 2.2).

Can we actually quantify the shrinkage of ginseng plants somehow? As it turns out, we can, by looking directly at plants from the past. An herbarium is a place where plant specimens are archived for later study by taxonomists, evolutionary biologists, and more recently ecologists. Each pressed specimen is labeled by date and location of collection. We knew that botanists had been collecting ginseng specimens for a long time at our herbarium at West Virginia University, so we decided to just start measuring them to see if these pressed plants had actually gotten smaller over time. To expand our study, we requested dated ginseng specimens from several other herbaria as well.

What we found was intriguing. In several dimensions, including leaf size and stalk height, plants were significantly smaller in the year 2000 than they were in 1900. Although we were excited by these findings, we had to acknowledge that even the conclusion that plants had shrunk made several assumptions. For example, we had to assume that botanists were collecting somewhat randomly at both times and therefore these plants were representative of the natural populations present then and now. Alternatively, their collection patterns had to be equally biased. In other words, we had to assume that botanists of the past were not more inclined to choose larger plants than botanists today. Also, most herbarium specimens had no roots attached (fancy that!), so it was not possible to compare root size changes over time directly. We do have evidence, however, that aboveground plant dimensions are correlated with root size. If we therefore assume that the proportional reduction in aboveground

FIGURE 2.2.
Left: Ginseng merchant next to shipment of wild ginseng roots in Buckhannon, West Virginia, in 1929. Right: recent photo of barrel of wild ginseng roots being inspected by USDA-APHIS. Scales of two photos approximately equivalent, based on shoes in photos. (Photo by Pat Ford)

size is reflected in the same proportional reduction in root size, we can do a rough calculation about the true reduction of maximum number of plants harvested from the 1880s (ca. 400,000 pounds per year) to the 2000s (77,000 pounds per year).

From the original herbarium data, I observed a 25 percent reduction in mean height of the stalk between the year 1900 and the year 2000 across all specimens. Assuming there was also a 25 percent reduction in root mass, this means that relative annual harvest went from a high of sixty million plants to about fifteen million plants. If this harvest reduction reflects change in wild ginseng plant numbers in the woods, which is by no means certain, that loss is quite dramatic.

Herbarium records have even been used to try to get indirect estimates of past abundance changes. This method might seem to have all the flaws of using harvest records. But Martha Case from the College of William & Mary cleverly compared ginseng collection rates with nonharvested, closely related species. This provided a kind of control, because if the number of ginseng herbarium specimens dropped, but so did the relative numbers of specimens of related taxa, there would be no evidence that ginseng was getting harder to collect and therefore scarcer. She found that there were proportionally fewer ginseng plants collected, particularly in the northern part of the range, suggesting again that on the century time scale, natural ginseng populations were declining. The lack of evidence for a difference in collection rate of ginseng and its relatives in the central and southern states did not necessarily imply that natural populations there were not declining. Changing and perhaps regionally different attitudes toward collecting rare plants could have obscured chronological trends in the data. Again, the indirect nature of inferences being made was a real problem, even while the science was sound.

How Much Ginseng Is There?

State botanists attach rankings to rare and endangered species in an attempt to prioritize conservation. Most often these rankings are supposed to reflect the number of populations in the state. However, as suggested by its broad geographic range, ginseng is able to grow on a wide variety of soils, aspects, elevations, and forest community types, so estimating numbers of populations is challenging. Early in our ginseng studies, we were foolish enough to make the attempt. Over two summers, with a large field crew, we thoroughly sampled forest understories quasi-randomly over seventy-two acres scattered over forests in much of West Virginia. Making a wild leap of mathematical extrapolation, we calculated the number of ginseng plants in the entire state.*

The magic number was 87.8 million! This number might seem absurdly large at first, but based on recent harvest numbers it suggested a harvest rate of 5 percent annually, so the number might not have been too far off. Using several sources about average population sizes, this number also suggests there are around 1 million populations in this one state. Certainly these numbers are only approximate, and the accurate numbers could be 25 percent higher or lower. But this contrasts dramatically with the state's ranking of the species as S3S4 (uncommon–rare; estimated to be 20–500 populations).

Why was the state ranking for ginseng so far off our estimate? We think it is because ginseng has a rare kind of rarity, and, because of that, botanists underestimate ginseng abundance. Most rare species are confined to small, specialized habitats, often habitats that are being lost. But ginseng can live in a broad spectrum of forest communities and is geographically very widespread. However almost all populations are very small, so each of them is vulnerable to extirpation. These small, infrequent, but widespread populations can give the impression of global rarity, but, summed up, there are still a lot of plants out there.

*We sampled only forests, on public land or private land where we received permission, and we "stratified" the random sampling such that we sampled a variety of elevations and aspects on purpose. We also avoided rhododendron thickets because we never saw ginseng under them, and for practical reasons—they don't call them "rhododendron hells" for nothing!

I have thrown around a lot of numbers in this chapter. But you should definitely feel uncomfortable with the high level of uncertainty concerning the answer to our question "Has ginseng declined?" and the even more important question "Is ginseng still declining?" Despite this, from the evidence presented so far, we can conclude that ginseng was much more abundant in the 1700s and early 1800s than it is today. This trend does not bode well for the ongoing preservation of ginseng or for continuation of the Art and Ellie story in Appalachia. Before jeopardizing that precious meme of a species being carefully preserved by cross-generational, conservation-minded families, conservation scientists needed hard data on real populations. We needed to know much more about the numerical changes in wild ginseng. But was this possible to do for natural plant populations? In the mid-1970s you could scour the scientific literature and find precious few examples. But then along came a proper, creative, wicked-smart British scientist who lived at the following address in North Wales: Glan-y-Coed Park, Dwygyfylchi, Penmaenmawr, Gwynedd. I kid you not.

CHAPTER 3

John Harper's Dream Plant

> In many ways plants make better material for the study of populations—plants stand still to be counted and do not have to be trapped, shot, chased or estimated.—**JOHN HARPER**

The Discipline of Plant Population Biology

In the mid-1970s, right around the time that Art Logan and granddaughter Ellie were exulting in their first ginseng harvest together, John L. Harper, an ocean away in England, almost single-handedly invented the field of plant population biology. Indeed, his seminal book *Population Biology of Plants* launched a generation of graduate students into the hottest new field in botany. Eager disciples of Harper in Europe and America fanned out across field and forest, tagging plants and documenting their dynamics in a frenzy to understand plant demography (the study of births and deaths in populations), competition, reproduction, disease effects, pollination, natural selection, and a host of other topics that had previously been largely the academic territory of zoologists. Scholars from across the world would make the pilgrimage to Prof. Harper's lab in Bangor, Wales, to sit at the feet of the master and absorb his wisdom, then return to their homelands, inspired to tag and measure and map their botanical subjects for season after season.

In the fullness of time, they discovered that Harper was mostly right. The sedentary nature of plants allowed ecologists to perform not only purely descriptive studies but also careful experiments entailing everything from controlled growth chambers to the variable environments of the field. And at the end of such experiments, no one cared if the plants were harvested, chopped into bits corresponding to different functions (e.g., roots, stems, leaves, and reproductive parts), and thrown into a drying oven for days, after which their dried carcasses were weighed to the tenth of a milligram on a lab balance.

Predictably for rapidly expanding fields, not only were the advantages of the new approach clarified, eventually the limitations came into focus as well. New techniques were developed to circumvent the limitations, and then new challenges arose in an ongoing process typical of maturation of any discipline.

One early bugaboo that quickly came to the fore was the challenge of counting seeds. With fauna, you could readily determine clutch size by counting eggs or chicks in the nest of a wood thrush, for example, or you could count the number of fawns trailing behind a doe, but seeds were often tiny and superabundant. Most notorious are orchid seeds, which resemble grains of dust in size, number, and dispersal. A single dandelion seed head contains fifty to two hundred seeds, and one plant can produce several heads a week for an entire summer. Oh, and seeds, which hold embryonic plantlets, do not "stand still and wait to be counted!" They move. A lot! Dispersal in space was a problem for researchers.

A second major challenge was that, in most plants, the largest number of individuals were out of sight! Invisible plants? No. This was a seed problem as well. As it turns out, many plants hedge their bets by producing seeds that do not germinate right away. They disperse in time, as well as space. In a lot of them, these dormant seeds hang around in the soil for years, decades, centuries, and in a few extreme cases, millennia! If you are trying to understand how numbers of individuals in a plant population change over time, you just can't ignore the fact that most of them are hanging around as a "seed bank" in the soil.

The theory is that dormancy evolved to increase the chance that seeds will germinate in the future at a time when they are more likely to be successful. Also, not being able to predict the optimum future point in time, plants were hedging their bets by spreading out germination over many years. Thus, dispersal in time results in this large cryptic component of a plant population. Unlike counting their postgermination counterparts, keeping track of this long-lived invisible fraction of the population was tricky.

A third fundamental problem came as a surprise to those used to working with unitary animals, and that is what to count as an individual. Defining an individual as the plant that arose from a single seed makes sense from a genetic standpoint. But as they grow, a great many plant species can splinter into semi-autonomous or fully autonomous sub-plants by vegetative propagation. Grasses, for instance, typically spread by tillering—a process that multiplies the growing tips of the shoots, followed by rooting of the new tiller, which may in turn produce more tillers, an insidious process that can eventually cover your lawn with crabgrass. What nonbotanists are surprised to discover is just how common vegetative spreading is. The largest plant in the world—indeed, perhaps the largest organism in the world—is a single aspen tree that has spread clonally by sprouting from underground roots. The Pando clone in Fishlake National Forest in Utah has spread to cover more than one hundred acres, weighs approximately fourteen million pounds, and is about eighty thousand years old. Looking at it, you would never know that the forest of tree trunks was actually just one individual. Genetic testing has confirmed that the entire clone arose from a single tiny seed.

To be fair, John Harper knew about all of these flies in the ointment of plant population biology and devoted many pages of his 892-page book to them. He simply challenged fellow scientists to study them and overcome them. Population biologists who stuck with it often settled on model species that conveniently had few of these problems. An example was jewelweed—a short-lived annual plant, each stem of which arose from a single seed. It had no soil seed bank, and it had large, readily counted seeds. Even if the seed pods did burst explosively when you touched them, the plant was a jewel to study.

Life History of Ginseng

I recite the following brief history of my chosen field because ginseng, as it turned out, was also a population biologist's dream, fortuitously for me. Vegetative cloning was rare (though infrequently a second stem would arise from a single root), seeds were large and easily counted as bulges on the fruits (only one-, two-, or three-seeded fruits were observed), fruits were easily counted (wild plants with more than a couple dozen berries were rare), and, as it turned out, while a seed bank existed, it was short-lived. In addition, most seeds appeared to just drop to the ground, so following their movements over large distances was not necessary.*

To understand ginseng's life history, the easiest place to begin is at the beginning. Or maybe even a bit before the beginning. Before life as a plant even begins as an embryo within a seed, the sperm of a male plant has to reach the egg of a female plant. In ginseng, as in a lot of plants, the sperm and egg are produced on the same individual. There are no males or females, just hermaphrodites. In fact, ginseng, like many plants, can even fertilize itself, in the most extreme form of inbreeding. Pollen carrying the sperm cell gets transferred from

*This turned out not to be true, strictly speaking, as some animal dispersal has been detected. More on that later.

the anther of the flower to the sticky stigma of the same or different flower on the same plant. Then the pollen sprouts a tube that grows down into the stigma, through the style, and into the ovary, where it delivers the sperm to the egg. In other words, ginseng can mate with itself and produce perfectly viable seeds.

Of course, a given ginseng plant can also mate with other ginseng plants, but this requires a vector—something to carry the pollen from the anther of the flower on one plant to the stigma of a flower on a neighboring plant. In ginseng, a few insects have been observed to carry out this function. Syrphid flies and halictid bees are the most common. These small-bodied pollinators match the tiny ginseng flowers. But they are not usually abundant in the woods, and they may or may not successfully pick up and drop pollen, even if they do visit. On top of that, visitation does not ensure mating between individual plants because the insect may move between flowers on the same plant, facilitating selfing. The bottom line is that sometimes seeds are produced by outcrossing and sometimes they are produced by selfing.

A detailed look shows that the story of ginseng pollination is even more intricate than I have described thus far. This is because plants tend to grow in small patches, and these patches are often made up of close relatives because seed dispersal is so limited. Often within a patch there is a large adult surrounded by medium adults, juveniles, and seedlings, suggesting a family grouping that has developed over time. Indeed, when the genetics have been examined for these groups, they are often closely related parents and siblings. So when an insect pollinator visits one adult, even if it carries pollen to a nearby adult, it may nevertheless facilitate inbreeding among family members. Only when the insect flies a long distance between patches does it aid in crossing between nonrelatives.

On the face of it, this would seem like a bad mating system. In humans, we have cultural taboos, often coded into law, to prevent inbreeding because mating among close relatives can result in inviable, or unhealthy, offspring. In biology, this negative effect of inbreeding is called inbreeding depression. This is not generally the case among plant species with a long history of selfing. The unhealthy genes that would otherwise come together due to common ancestry when mating occurs between relatives have long ago been purged from the population. However, a further complication with ginseng is that the high degree of selfing seen now may be, in the big scheme of things, a relatively new occurrence. Why? Because, as described in chapter 2, populations in the past were much larger and denser than they are today. Those larger populations of three or more centuries ago would likely have placed nonrelatives in closer proximity to each other than we see with the patchy family groups today.

Is there evidence of inbreeding depression in ginseng? As it turns out, there is, although the jury is out concerning how severe it is. In her PhD research on ginseng, Emily Mooney assumed the role of insect pollinator, bagging flowers and pollinating either within or between individuals to experimentally create seeds that resulted from selfing or outcrossing within a population. She then planted those seeds, waited for them to germinate, and followed them for two years. By the second year, progeny produced by selfing were 31 percent smaller than those produced by outcrossing. Given that size is related to survival, this was taken as evidence for inbreeding depression.

Regardless of genetics, once the egg is successfully fertilized in a newly forming seed, the young ginseng embryo can begin to form right there inside the developing seed. The seed forms inside a fleshy fruit called a "berry," for lack of a better term. Ginseng berries don't contain as many seeds as other fruits that we call berries, such as strawberries or raspberries. But the structure doesn't fit the category of a peach or a cherry, which is called a drupe, because it lacks the pit. Ginseng fruits start out green and contain one, two, or more rarely three seeds. As long as the plant has sufficient water, nutrients, and sugars to develop the berry, the fleshy berry containing the seeds will grow to about the size of a pea, though they are shaped more like a kidney bean.

If you sliced open a very young seed, you would have a very difficult time seeing the embryo without a microscope. But after a year or two lying in the soil, the embryo has grown and begun to absorb the starches and nutrients held inside the seed coat. Finally, after two winters (sometimes three, four, and rarely five—more on that in chapter 5), the young stalk bursts forth out of the seed and two young leaves unfurl. A new seedling is born. Wait—are there no "seed leaves," or cotyledons? As gardeners know, most plants have these different-looking leaves right after germination, but ginseng has retained them inside the seed and skips the step of producing cotyledons aboveground.

The new seedlings of ginseng face daunting odds of becoming established that first year. Fortunately, unlike tiny orchid or dandelion seeds, they start out fairly well-endowed by their mothers with carbon and nutrients to grow. Seedlings start out with a single leaf with three leaflets, and in that first year, the middle leaflet tends to be long and narrow. In addition, the stalk under the leaflets lacks a distinctive bump that will be present in future years. The third diagnostic of a new seedling is that the seed coat may still be attached just below the soil surface along the stalk. Later on, that seed coat will fall off and disintegrate in the soil. The key to success of the new seedling is establishing the root. A root that grows down and discovers a reliable source of moisture and nutrients will likely mean survival for the young plant as a whole. Otherwise one will die.

For most perennial plants, in year one after germination the fate of death is far more common than survival. This is true for everything from redwood trees to dandelions. One of the strangest things about ginseng is that this is *not* true. Survival of ginseng seedlings over their first year averages about 75 percent. That is truly staggering.

What accounts for ginseng's high seedling success rate? No one really knows, but we can form reasonable hypotheses based on a few facts. Their relatively large seed size means that the first few weeks of growth can be sustained by the reserves in the seed itself until the first leaflets appear above the leaf litter, and the main rootlet heading down can reach more compacted, water-holding soil. Next, the forest understory is protected from harsh sunlight by the tree canopy and larger herbaceous plants above, so seedlings are unlikely to lose a lot of water through leaf pores known as stomates. The eastern deciduous forest is also known for its even distribution of rainfall during the growing season, again ensuring that the young plants do not die from drought stress.

In year two and beyond, a ginseng plant may simply remain a seedling, defined as a plant with a single leaf (with three, four, or five leaflets), for up to five years or more, depending on environmental conditions. These are the toddlers of the ginseng population. For example, in one population, admittedly in marginal ginseng habitat, 58 percent of five-year-olds still had only one leaf, appearing stuck in this stage. One plant in this population was still a one-leaf seedling after fifteen years! On the other hand, on a good site not too far away, less than 35 percent of the five-year-olds were one-leaf plants. This illustrates a fundamental principle of plant growth—its tremendous plasticity. You can see this in your garden if you fertilize some plants but miss others. In the wild, these differences are exaggerated. For ginseng, after the leaves stop expanding in the spring, there will be no more change that year; size changes aboveground show up the following year.

What are these slow-growing plants waiting for? Time may be all that is needed for the little ginseng plant to accumulate enough energy to grow to a larger size. But often it appears that a pulse of energy is needed to boost them into the next size class. We observed many times that remarkable size changes would occur if the overstory tree canopy was slightly opened up. Lightning strikes, insect defoliation of the overstory trees, windthrow due to hurricanes or derecho events, ice storms—all of these perturbations, if not too severe, could stimulate growth of ginseng in the understory. Thus, while it is very much an understory plant, it is possible, especially for a small one, to receive too much shade.

When a one-leaf plant graduates to a larger size, this shows up aboveground when it adds a second leaf. Ginseng aficionados refer to leaves as prongs, so a two-leaf plant would be a two-prong plant or "two-pronger." These are the children and adolescents of the ginseng population. As a catchall phrase for two-leaf plants, we refer to them as juveniles. Quite often, juveniles will sprout a reproductive structure at the junction where the two prongs meet. The structure can be a full-blown inflorescence (the cluster of flowers at the top of the stalk), but typically they are small compared with those of adults, having just a few flower buds. These flowers produce stamens, anthers, and pollen quite often, so they could act as fathers, but juveniles act as mothers (produce seeds) only infrequently in the wild, likely because this requires a much greater investment in energy than needed to produce the tiny pollen grains. It is not uncommon for juveniles to abort the entire inflorescence, presumably saving energy for storage and root growth. During juvenile development, the central root begins to thicken and lengthen, a bit more each year.

After a few more years (around three to ten), the two-prong plant may graduate to adult status, with three or more leaves. Adults produce an inflorescence almost every year but may or may not produce seeds. Seed production is partly a function of internal resource accumulation and partly determined by activity of pollinators. Plant population biologists discourage equating plant life histories with humans—I plead guilty of exactly that in my description so far—and, indeed, analogies with human life stages break down in the adult class. Adult humans stop growing at some point, at least in basic bodily dimensions such as height. Plants, on the other hand, are quite plastic in size, and though they stop growing after leaf expansion within the course of one year, the following year they can come back a different size. Oddly, however, plants can get not just larger but also smaller. An adult ginseng can continue to grow larger by having bigger leaves and a taller stalk, but it can come back smaller the next year if some stressor comes along. For example, if intense drought occurs in July and August of a given year, the top can be lost to early senescence, and the top will reemerge smaller the following year. An adult three-pronger can even come back as a two-pronger, for example, and this is not uncommon. On the other hand, in a good year, energy may be accumulated as stored reserves in the root, and a new, larger top will be produced. A three-pronger can come back the next year as a four-pronger, with even larger leaf area. The stages in a ginseng life cycle are summarized in figure 3.1.

FIGURE 3.1.
Summary of stages of growth. Progression through the stages is not steadily forward; plants may get larger, smaller, or stay the same from one year to the next. Sizes and size changes reflect energy storage from the preceding year or years.

All of this shrinking and growing from one year to the next matters because of a fundamental principle of plant populations that became clear very early in the field's development: reproduction, survival, and even growth of individuals are size-dependent. So, while animal population biologists scurried around measuring age-dependent survival, growth, and reproduction, plant population biologists quickly switched to measuring these same statistics as a function of size. This solved two problems for the plant folks. First, often age was difficult to determine for an herbaceous plant, particularly one that reproduced by vegetative spreading. Size was much easier to measure. And second, since plants were so plastic in their response to the environment, size better captured what mattered in terms of survival, growth, and reproduction.

The beauty of ginseng as a subject for plant population biologists was that measuring its size was a simple matter, accomplished with a ruler. Once leaves had stopped expanding in mid- to late May, the plant's aboveground stalk and leaves were fixed in size. Changes in size could therefore be assessed through measurements made once per year. On ginseng, some researchers counted prongs. We chose to measure the central leaflet on each leaf because this set of measurements was a great predictor of the leaf area of that leaf. We could then add together the leaf area on all leaves to get the total leaf area of the individual. We also measured stalk height, which was highly correlated with leaf area.

Flies in the Ointment?

No plant is perfect for demographic study. Ginseng had two aspects of its life history that presented challenges. First was seed dormancy. We already knew that seeds would lie dormant for two winters before germinating. However, early on, we discovered that not all seedlings emerged after twenty-one months, so we had to study them as a separate group from the aboveground plants and quantify that variation.

The second tricky thing for ginseng was the occasional plant that was present in year one, went missing in year two, but mysteriously returned in year three, very much alive. Given that roots cannot get up and walk away, what was happening with these temporarily missing stalks and leaves? This mystery had vexed ginseng lovers for a long time. Indeed, fantastical stories were invented to explain the ability of ginseng to hide from seekers. Some stories implied that a ginseng hunter must be of both high and humble character and worthy of the plant's favor before the plant would reveal itself. Nothing so fanciful was the actual reason, but it took a real forensic sleuth to solve the mystery.

In the meantime, with its unitary structure, its easily counted short-lived seeds, its slow growth, and its easily reached and measured stalk and leaves, ginseng was indeed a plant population biologist's dream. In following populations of one plant over time, we had a real chance of not only detecting change (the subject of the following chapter) but also of making predictions for the future (the subject of chapter 5). And that meant we had the possibility of determining whether species protections put in place with CITES were adequate to the task or still lacking the ability to save this precious resource.

CHAPTER 4

A Species in Decline?

If we have data, let's look at data. If all we have are opinions, let's go with mine.—**JIM BARKSDALE**

Sampling: The Real and the Ideal

The field crew, consisting of four undergraduates, two graduate students, and one professor, was getting tired and frustrated. We were in the middle of day two of prospecting for wild ginseng populations to study in a West Virginia state forest that will remain unnamed to protect the location. It was July, and the weather was hot and humid. Wearing long pants and long-sleeved shirts, we had sweated our way through acres of forest understory consisting mostly of poison ivy and thorny greenbrier, ducking under spicebush branches and keeping our eyes peeled for timber rattlesnakes. We had found exactly zero ginseng in almost two days of hard searching. Our admiration for wild harvesters only increased, imagining them spending their precious fall weekends and, like us, finding nothing for their trouble.

The lunch break couldn't come soon enough. We sat down and, over sandwiches, Pringles, and long sips of water, wondered aloud whether we were using the right approach to find our quarry. Nearing the end of our break, Hannah, one of the undergraduates, declared that she needed to go in search of a private tree. A couple of minutes later the rest of us heard a shriek, and we quickly realized something must have gone very wrong. However, the next thing we heard was "Ginseng! I found ginseng! I almost peed on it!" Everyone broke out laughing, realizing all at once that Hannah had singlehandedly discovered a new way of locating ginseng. For the rest of the summer, we called it the "Hannah method," and, believe it or not, it worked a few more times! Returning to our rigorous, more systematic, and scientific approach, we spent the rest of the day flagging one of the largest populations we ever found for future monitoring.

How do we know a species is endangered or threatened? Fundamentally, this is a problem concerned with abundance (population size, often given the symbol N) and changes in that statistic over time. The most straightforward way to show whether populations are declining would be to find a randomly selected array of populations across the range of the species, then monitor changes in N over time. In surveying the West Virginia state forest, the random method is exactly what we were attempting.

Population monitoring can be surprisingly hard to do. First, a "population" needs to be defined. This is not trivial—imagine doing this for a human population arrayed across the landscape. Where does one population end and the next one begin? Biologically, the definition of the population should center on migration of genes among and within groups; populations are groups that are relatively isolated from other such groups, but following genes around the landscape is challenging. Second, true random sampling is much more difficult than it seems, particularly for rare or uncommon species. Ideally, one would locate, say, ten thousand natural populations by randomly setting up plots and censusing them. Then you could use a random number generator to pick a representative sample, of say, one hundred of them to monitor over time. With a rare plant, you might have to set up and survey a million plots to find ten thousand populations, so clearly this would be a prohibitive amount of work.

Continuing our ideal scenario, after finding one hundred populations to monitor, you would need to understand the within-year dynamics of the population so you could census them at the time of year when you are most likely to count all the individuals. With our hard-won knowledge of ginseng's life history, we were ready to take on a more practical version of this challenging task.

Defining a Set of Central Range Study Populations

For ginseng, we ultimately determined that the ideal scenario of purely random sampling was simply impractical without gargantuan amounts of funding to pay armies of trained field help. Therefore, we decided to content ourselves with finding "representative" populations of wild ginseng to study that would be "close enough" to random. Some of these populations were randomly discovered through surveys, but others were found through recommendations of private landowners who were known to university extension agents or others to have ginseng on their property. We were alerted to other populations by park rangers or other kinds of land managers. We therefore do not claim

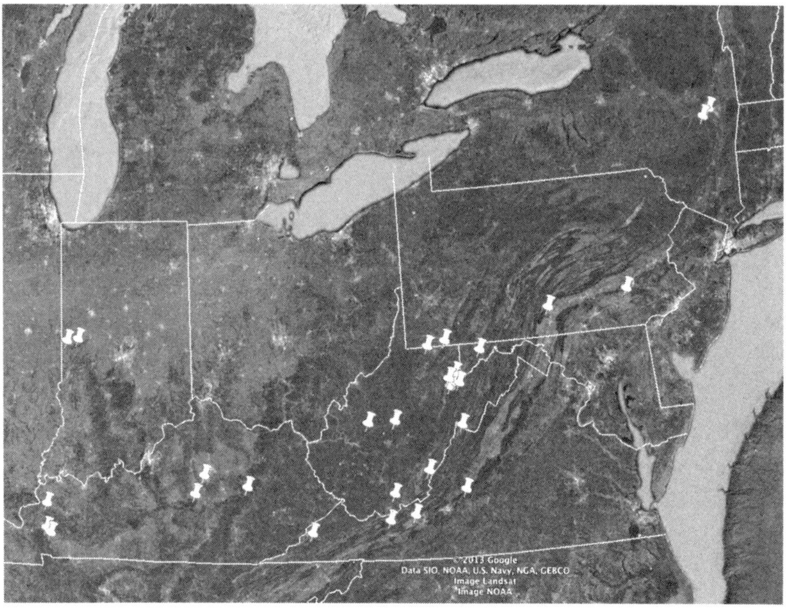

FIGURE 4.1.
Google Earth image showing the distribution and rough locations of thirty monitored populations of American ginseng; all were censused annually in spring and fall for thirteen to nineteen years each

these are a completely unbiased set of ginseng populations. But these populations had to meet certain criteria before we would formally include them in the study:

1. **We wanted representation of populations over much of the central range of ginseng.** By 1998, we had identified several natural populations of ginseng by way of random sampling, but this process was painfully slow. Therefore, over the next six years we used our other contacts to assemble a group of thirty populations spread over seven states: Indiana (2), Kentucky (6), Maryland (1), New York (2), Pennsylvania (2), Virginia (5), and West Virginia (12); see figure 4.1. Clearly this sample is biased toward West Virginia, Kentucky, and Virginia for practical reasons—most of these sites were within a day's drive of our base in Morgantown, the home of West Virginia University. However, one justification for the Appalachian bias was that four of the top five wild harvest states each year are located there (fig. 4.2). Missing from this sample are populations from the periphery of ginseng's range: the southernmost states (especially Tennessee, Georgia, North Caro-

FIGURE 4.2. Average ginseng harvest per year (2004–2013) from states where harvest is allowed

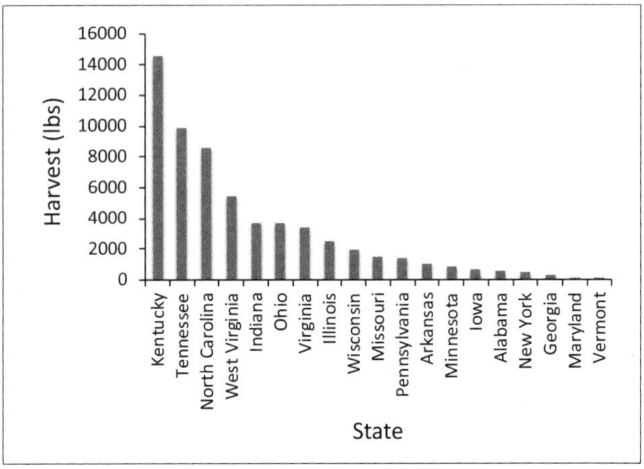

lina, Arkansas, and Missouri), the northwestern states (Wisconsin, Minnesota, and Michigan), and the Canadian provinces (Ontario and Quebec), with apologies to Ohio for skipping you.

2. **Populations had to meet the criterion of "wild," as defined by CITES.** In practice, this restriction meant that (a) we had evidence that the populations were not originally planted by the landowner, and (b) the environment was not specifically manipulated to encourage ginseng growth (no fertilizers, pesticides, irrigation, soil amendments, forest thinning, etc.). In fact, ginseng buyers have no guarantee that a root being sold is "wild" and not from "wild-simulated" (i.e., planted) populations. However, our long-term monitoring helped assure us that that the above requirements were being met. This did not mean that no harvest occurred in these populations—indeed, we were specifically interested in the rate of wild harvest, so this was noted wherever possible. The term "wild" is controversial in the ginseng harvester and grower community because of the frequency and long history of interactions between people and ginseng. "Untended" or "natural" populations might be better descriptors, but for reasons of precedent as well as common use, I will continue to use the term "wild."
3. **Populations were selected from a wide array of elevations, aspects, land use histories, and overstory tree community types.** Without a truly random sampling approach, we cannot claim these thirty populations were representative of all wild populations, but by sampling a mix of low-, medium-, and high-elevation sites, and making sure we sampled private lands as well as public lands, we could see if these factors affected population viability. Table 4.1 summarizes characteristics of the thirty populations used in this

TABLE 4.1.

Summary characteristics of thirty censused populations of ginseng (1998–2016) across seven eastern U.S. states

Population*	State	Census Years†	Elevation (feet)	Aspect	Harvest History‡	Browse (%)§	Overstory**	Ownership††
1	IN	14	659	SE	Yes	18.5	MM	SP
2	IN	17	607	E	No	23.1	MM	MR
3	KY	14	997	NW	Yes	6.2	MM	PP
4	KY	15	492	N	No	13.4	MM	PR
5	KY	15	909	N	Yes	1.6	MM	PR
6	KY	16	827	W	Yes	4.7	MM	NC
7	KY	15	397	SW	No	31.9	MM	PR
8	KY	15	499	SE	No	11.7	MM	PR
9	MD	13	1,854	N	Yes	8.7	MM	NC
10	NY	13	1,539	S	Yes	24.8	AH	PP
11	NY	13	1,394	N	No	15.5	AH	SP
12	PA	13	735	S	No	14.9	MM	PR
13	PA	13	797	E	No	20.6	MM	MR
14	VA	16	2,175	NW	Yes	28.3	MM	PR
15	VA	16	3,504	NW	No	67.0	AH	PR
16	VA	16	2,503	W	No	40.0	OH	NP
17	VA	13	2,005	N	No	25.1	MM	MR
18	VA	16	2,979	NW	No	24.6	OH	NF
19	WV	16	2,507	E	Yes	18.8	MM	MU
20	WV	17	2,467	NE	No	29.7	AH	PR
21	WV	17	2,480	S	No	4.9	AH	NF
22	WV	16	1,181	E	Yes	18.3	MM	PR
23	WV	17	2,133	E	No	14.8	AH	NF
24	WV	13	1,965	NE	No	33.3	MM	PR
25	WV	17	942	N	No	19.5	MM	PR
26	WV	17	1,545	NE	Yes	11.7	MM	SF
27	WV	19	1,676	N	Yes	20.7	MM	SF
28	WV	16	1,129	E	No	54.0	MM	PR
29	WV	17	2,585	N	No	11.2	AH	NF
30	WV	16	2,641	SE	No	17.5	AH	NF
AVERAGE		15.37	1,604		36.7% Yes	21.2		

* Population names and precise locations have been obscured to protect them from additional harvest pressure.

† Number of continuous years of formal demographic censusing; spring and fall census every year

‡ "Yes" means harvest occurred at some time during the years the population was censused.

§ Browse: mean percentage across ten census years (2004–2013)

** MM: mixed mesophytic forest; AH: Allegheny hardwoods; OH: oak-hickory

†† MR: military reserve; MU: municipal; NC: nature conservancy; NF: national forest; NP: national park; PP: private preserve; PR: private property; SF: state forest; SP: state park

study. Three populations, noted with an MR under "Ownership," were located on protected military reservations, and these served as a kind of "control" with a long history of no harvest.

4. **Populations were defined pragmatically as well as biologically.** We defined a population as any grouping of plants that was at least one hundred meters (a bit more than a football field) away from any other such grouping. In most cases, we tried to find and mark every individual in the population, although in a couple of cases, the populations seemed to extend so far that censusing all of them would have been impractical. All were located in plots of land less than ten acres in size, and most were much smaller than that. The distance of one hundred meters to separate populations was based on observations of pollinator movements (small bees and flies) as well as an early experiment in which we showed a falloff in seed production around fifty meters away from a central pollen source (see chapter 13). The third piece of "gene flow" information came from recent studies of bird dispersal, showing that seeds would be dispersed as far as one hundred meters only extremely rarely by the main bird dispersers of ginseng seeds (see also chapter 13). While detailed genetic analyses could be done to define populations even more precisely, these analyses have not been done. Despite this, we know more about dispersal distances of pollen and seeds for ginseng than for most plants, so the hundred-meter criterion seems reasonable.

Population Boundaries

We wanted to census our thirty populations every year to accumulate data about every individual. But first we had to find every plant. You might think this would be easy. After all, as John Harper exclaimed, plants "sit still and wait to be counted." In practice, this is very challenging because sometimes the vegetation is dense, and picking out a single small ginseng plant in a tangle of other plants is tricky. This is especially true for the smaller one-leaf plants that are often hidden. While ginseng is distinctive, especially when in flower or bearing fruit, you would be surprised at the number of lookalikes that draw the eye. Several hours of scanning for ginseng taxes the brain, and many a conservation intern complained of seeing ginseng in their sleep as they processed all they had seen in the course of a day establishing a population.

Delineating a population was done with teams of five or six trained ginseng spotters (undergraduate and graduate students) arrayed in a line with about fifteen feet between each person, then slowly walking in tandem across a slope, marking every ginseng plant with pin flags. When we had gone about a football field without seeing a plant, the line of spotters reversed course and walked a new parallel transect back across the population, flagging plants. This process was repeated until we were convinced we had flagged every visible plant in the population. The team approach often turned into a good-natured competition between students to see who could spot plants first and ultimately flag the most individuals.

Locating Plants Again Year after Year

After flagging the visible plants in a population, we had to devise a method for finding them again more quickly the next time we visited. We could not redo the entire discovery process each year; that would have taken too much time. And we couldn't leave the flags in place, as this would have enabled would-be harvesters to easily wipe out an entire population. The invention and spread of digital photography during the early years of our research made the taking and printing of photographs very cheap, so we devised a method using "phototrails." Starting at a prominent visible and semipermanent object (usually a rock or tree) on one edge of the population, we took a compass bearing toward the first flag, captured an image of the path, and paced out a rough distance to the flag. A second image was taken six to eight feet from the flag, and often a third closeup image was taken showing the individual plant with its local surroundings. This process was repeated, proceeding from flag to flag, or from local fixed points to flags, through the entire population. Later these photos were printed and placed in order in binders with written instructions so when we returned later that year, or next season, we could retrace our steps, following the phototrail to each plant. After several years, phototrails would be rejuvenated with new photos when changes had occurred (e.g., decomposing logs disappeared or whole clusters of plants were absent).

The phototrail approach was combined with rough sketches of clusters of plants showing a few square feet of ground in greater detail. First, the locations of rocks, living tree trunks, down branches or trees, etc. were sketched on a piece of paper along with a rough scale and compass orientation. Then individual plants were mapped in, using a dot for a seedling, an x for a two-pronger, a triangle for a three-pronger, and a square for a four-pronger.

To avoid calling attention to our plants with a visible tag or flag, we found that aluminum roofing nails could serve as long-lasting markers. Unique identifying numbers were etched onto the flat top of the nail. A nail was placed at the base of each plant next to the rhizome where the stalk emerged each spring. As new seedlings were discovered, new numbers were assigned, and they were followed through time. When a plant died, its number was retired and not re-used, to keep future identifications clear. If we discovered a new plant that was somehow missed in the initial census, we added it to the population and to the phototrail so we could have a complete accounting.

By 2004, with our thirty populations all circumscribed (table 4.1), we were censusing 4,357 plants twice annually. Gradually, as we added new seedlings and new plants that we had missed in early censuses, this number increased to 5,242 plants by 2010, but the number declined after that year. All thirty populations were censused completely every year until 2016. In total, we collected data on 13,841 unique individuals over that span. Other plant population studies have included more plants, and a few have even gone longer. But, to our knowledge, this study has the largest sample of combined "population-years" (461) of *any* natural plant population in which a complete demographic census was performed.

The Census

Natural plant populations are always changing. To know how many individuals are present, you have to select a point in time when the highest number of plants will be visible. Knowing the timing of events in a plant's life history, its phenology, is critical. Although the timing of events will vary geographically, and even year to year in a given population, the overall progression is predictable year to year. Figure 4.3 depicts the phenology of ginseng.

We determined that plants would be fully expanded by the third week in May in most years. At the same time, because it was early in the season, few plants would be browsed by deer or damaged by other animals or events prior to that time. Also, new seedlings generally had emerged by then. Therefore, we censused the plants starting in the fourth week of May. By then, the plants would have stopped growing, but we could census almost all of the plants in the population. We started with the lower elevation, western and southern populations, and progressed to higher elevations and more northern populations as the census continued.

Each spring census was carried out by two field crews of three people, every year of the study. We measured or noted the following attributes of each plant in the spring census: (1) presence, (2) leaf number, (3) leaflet arrangement of each leaf (ginseng leaves have one to seven leaflets), (4) height of the stalk from ground to the point where the leaves branch off from the stalk, (5) lengths and widths of the longest leaflet on each leaf (this was used to estimate leaf area later), (6) reproductive status (presence of inflorescence or not),

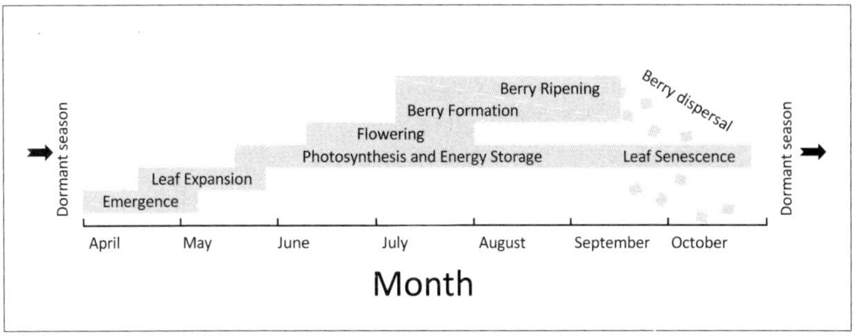

FIGURE 4.3.
Progression of annual events in the life of a ginseng plant. Within a population, emergence and senescence are often synchronized, though flowering, berry formation, and dispersal may occur over an extended period in a year. You can observe this progression in a single plant in a unique video on YouTube following a plant through 168 days in one year; search for "A Year in the Life of a Ginseng Plant."

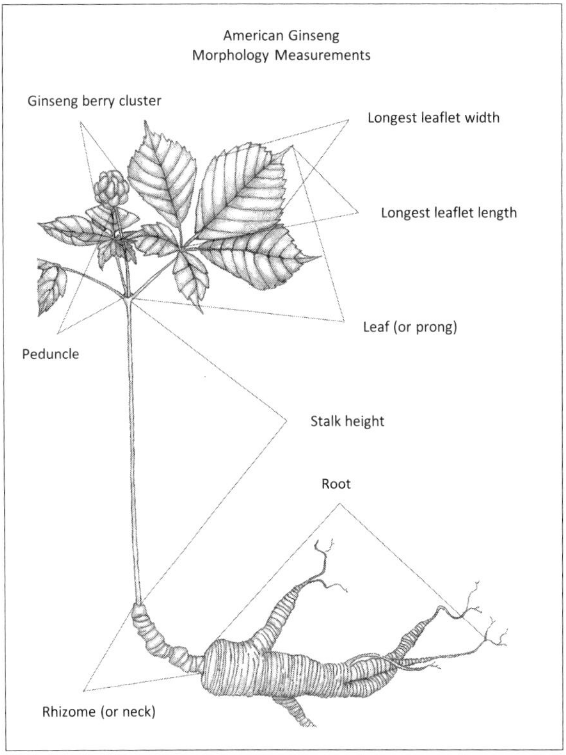

FIGURE 4.4.
Labeled parts of a ginseng plant that are measured

(7) deer browse status, (8) disease evidence, (9) insect damage, (10) frost damage, (11) and any other miscellaneous conditions or traits. The various dimensions measured are shown in figure 4.4. In addition, if the identifying nail was missing, we would note that the nail needed to be replaced. If a new plant or new seedling was found, it was added to the phototrail and to the data set.

The main purpose of the fall census was to count seeds. Therefore, we censused plants in the first two weeks of August, prior to dispersal and senescence but before harvest season began. As noted, ginseng berries can contain one, two, or three seeds, and it is easy to count them due to the distinct bulges caused by each seed within the berry. At the fall census, we also noted the color of each berry on the plant. In addition, we noted any new evidence of deer browse, disease, insect damage, harvest (which would have been illegal), or other traits of the plant.

Up or Down? How We Estimate Population Changes

Crunching all these numbers into a meaningful assessment of how natural ginseng populations are faring is no simple matter. Ultimately, we wanted to know whether the populations were increasing or declining over time. Population biologists seem to delight in making this calculation as complex as possible mathematically, but really it comes down to the answer to this question: Are there more deaths than births in the population? If so, the population is declining. If there are more births than deaths, the population is increasing. Because our censuses allowed us to follow both births and deaths, we could do this calculation.

But first, some nuances. What constitutes death in ginseng? Because plants can be browsed by deer, or damaged, and produce a new top the following year, we cannot judge death by the absence of the plant at one point in time. We have to return to the location for multiple years to determine for sure that the plant is dead. As it turns out, if new seedlings, or older one-leaf plants are gone for one season, we can be sure they are dead. However, plants with two or more leaves require two seasons of absence for us to be sure they are dead. This means we cannot be sure about death for our last two censuses (2015 and 2016), so we stopped with the 2014 census in our official population time lines, even though we did census beyond that point.

Another kind of death in ginseng is seed death. This is more difficult to determine because we cannot easily tag every seed and follow it in the soil. However, technically, seeds are individuals because they contain a developing embryo of a ginseng plant. A total count of individuals in the population really should include them. But, for practical purposes, we will for the time being simply count individuals aboveground—that is, after germination.

And what are "births" in ginseng? Literally, an individual is created when pollination is successful and an ovule or ovules begin to develop into a fruit with a seed containing a developing embryo. Again, practically speaking, if we are considering only the aboveground population, we could define "births" as new seedlings emerging each year. But we should always keep in mind that a seed population exists in the soil. Later, when we make more formal models of population growth (chapter 5), we will have to account for those pesky seeds and devise a method to account for their numbers.

I will avoid equations in this book. (You're welcome!) But if you are to have confidence in the numbers, you should know what we did. So let me describe how we can produce a simple graph showing how numbers are changing with time.

CALCULATIONS

We start in spring of year one with a certain number of plants that we have censused. We can estimate how well this population is doing over the next year by adding any new seedlings that have germinated prior to the next spring's census, then subtracting any deaths of plants we observed in the set of plants we censused between year one and year two. Here is a simple illustration. Let's say in spring 2004 we censused 42 plants in Population 1 from Indiana. Then in spring 2005 we counted 4 new seedlings, and between 2004 and 2005 2 of the original 42 had died (we would need data from 2006 and 2007 to determine that). We could infer that there would be 42 + 4 – 2, or 44 plants in 2005. So the population changed from 42 to 44.

Key Assumptions

Because determining whether natural populations are increasing or decreasing is so important (recall this is a key to understanding sustainability of harvest under CITES), we need to carefully consider the assumptions that go into this calculation. First, we are assuming it is OK to ignore the seeds as part of the population—again, technically, this is not a good assumption, but it is probably reasonable to assume that seeds represent a certain fraction of the total population, and that as long as this fraction is not changing wildly, we can at least know that the true total population will simply be a certain percent larger than our observed aboveground population. We'll have to live with that assumption for the time being.

Second, we are assuming that our missing plants (gone one year for seedlings, gone two years for all others) are truly dead. Rarely, we do see plants gone for two consecutive spring censuses (probably browsed by deer), and they do come back in the third year. However, we can retroactively go into our data set and change them to "alive, but dormant" and correct for this problem. It really only becomes a problem at the very end of our data set, when the rare plant missing in 2015 and 2016 might return in 2017 (and we wouldn't know it because censusing was halted). This could result in a very slight overestimate of deaths in that last year.

The third assumption, and perhaps the most important, is that we have accounted for all new seedlings. In looking for new seedlings each year, we were forced to do what was practical. The majority of seeds passively fall to the ground, and, when they do, they land very near the parent plant. But one ambitious and enterprising graduate student then asked, "Once a fruit has fallen to the ground, how far do its seeds move?" The details of this study are given in chapter 13 on dispersal, but for now I summarize the result. She found that 90 percent of the seeds stayed within six feet, seven inches (two meters) of the spot where they were dropped. Those that did move generally moved downhill, probably washing downhill in heavy rains. Knowing this, when we assessed new seedlings we included all that were found within two meters of the parent plant. This suggests when we add new seedlings to determine population change, we should increase our number by 11 percent (1/0.90) since we know we are underestimating. When applied in other populations, this makes another assumption—that seeds will move at the same rate there. Certainly for populations on flat ground, we would expect less movement, but on steeper slopes seeds might go farther. Again, the fate of seeds turns out to be the difficulty of precise plant population biology.

The complications go further. We now know that ginseng fruits are beloved by a group of birds called thrushes (see chapter 13 again for this story). Wood thrushes, in particular, have a range that overlaps almost precisely with ginseng, and they disperse ginseng seeds. What is more, we now know that 91 percent of thrush-dispersed seeds travel farther than two meters from the parent plant. We have estimated based on camera traps at six ginseng populations that up to 37 percent of ginseng fruits might be thrush-dispersed. When adjusted for mortality of seeds in the soil prior to germination, this means nearly 37 percent of new seedlings could be missed by our censuses if thrush dispersal is that common. We know from feeding trials that thrush-dispersed seeds are perfectly viable, and therefore we feel comfortable assuming their germination rates and early growth are not affected by having been swallowed by a bird.

Conservation biologists tend to be conservative, as their name implies, when making population forecasts. This is because it is better to err on the side of being too pessimistic than to be too optimistic when considering whether a species is endangered. In other words, we will lose fewer species if we take more immediate action than we will if we react too late based on overly optimistic projections.

To be conservative with our data, we can project population sizes assuming that (1) no seeds travel beyond two meters (i.e., our new seedling data are complete), (2) thrush dispersal is absent, and (3) plants designated "dead" after two years really are gone. All of these are pessimistic from the point of view of conservation and therefore would yield a "worst-case scenario." We have proceeded with those calculations but will label those projections "worst-case scenario." These projections might well be realistic for a population on flat ground (little secondary movement of seeds) with no thrushes resident near the population.

While conservation actions should be driven by conservative estimates, we can provide an estimate of the opposite—an optimistic scenario—in order to provide an alternative view. If the "optimistic scenario" still suggests the species is in trouble, then we can know for sure that improved conservation protections are warranted.

For our optimistic scenario, we adjusted new seedling numbers upward for the ones that would germinate on occasion farther than two meters from our censused plants; our figure was 10 percent. This effectively adjusts our new seedling numbers upward, knowing we may have missed any new seedlings beyond our two-meter census radius. We assumed an additional 10 percent of seeds were moved beyond two meters by bird dispersal. This projection might be realistic for a population on a moderate slope with a population of thrushes nearby. Note that this is also optimistic because it assumes seeds did not move, by gravity or by birds, to within two meters of another plant that we censused—we would have detected all of those.

As you can see, even when we are being as scientific about this as we can be, there are no easy answers. The best we can do is to try to lower the uncertainties, then make projections that range from possibly unrealistic, though conceivable, on the conservative end, to unrealistic but possible on the optimistic end of the spectrum. These two estimates provide brackets on reality. This may be unsatisfying, but it is the nature of ecology. One of my graduate school mentors, Dwight Billings, used to declare "You've gotta tolerate a little slop to be an ecologist!" and this was Exhibit A for that quip.

But acknowledging uncertainty is not all bad. If, despite incorporating the variation that could be there, we still see a common pattern (e.g., of decline or increase), then we can be confident in our conclusions. This is exactly what climate scientists do in making predictions about Earth's climate future. They cannot predict with certainty disruptive events such as volcanoes any better than we can predict whether thrushes will be present in a particular ginseng population. Luckily, or unluckily, in terms of climate change, the uncertainties do not change the prediction very much.

Drum Roll, Please

So, what did we find? The charts below illustrate the pessimistic and optimistic scenarios for our thirty populations from 2004 to 2014. In order to make the changes easier to view and less like a tangle of spaghetti, we have "relativized" all the initial population sizes to one hundred. The percent changes from there reflect the actual adjusted "birth" and "death" rates in the populations.

The first take-home message from tracing the fate of these wild ginseng populations over a decade is that there was a general downward trend (fig. 4.5). The "average" of the thirty populations declined by 30 percent over that time, or roughly 3 percent per year. Indeed, while there was some year-to-year variation in that decline, there was only one year that did not show a decline overall (2009 to 2010). If populations were holding steady, we would expect about fifteen of them would have increased and fifteen would have declined, but the actual numbers were five increasing and twenty-five declining. This difference in observed increases and decreases is statistically significant, meaning it was extremely unlikely that the greater proportion of declining populations was simply due to chance.

A second take-home message from the figure is that populations varied. A lot! The fastest-growing population increased 70 percent over the decade, and remember this was the pessimistic scenario. This suggests that while the species overall is doing very poorly, not all populations were doing poorly. So the problem with ginseng was not universal across its range. This is reason for optimism if we can figure out why a certain minority of populations are doing well, while the majority are doing poorly. Whether we have ginseng a century from now or lose it from the forest may rely on whether or not we solve this puzzle.

A third take-home message, also offering some optimism, is the fact that none of these populations has been extirpated (N reduced to zero) in the span of ten years. Indeed, we will take a closer look at four specific small popula-

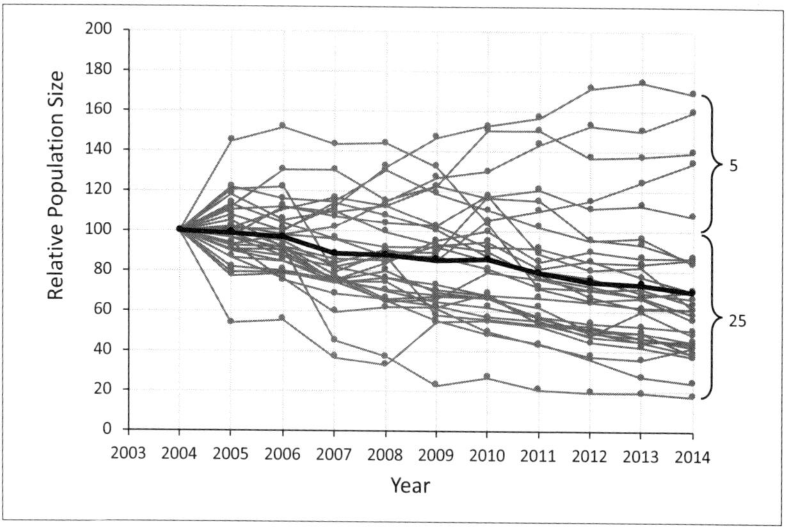

FIGURE 4.5.

Plot showing "pessimistic scenario" fates of thirty populations of ginseng over a decade, with all starting populations relativized to one hundred. The solid black line represents average of all populations.

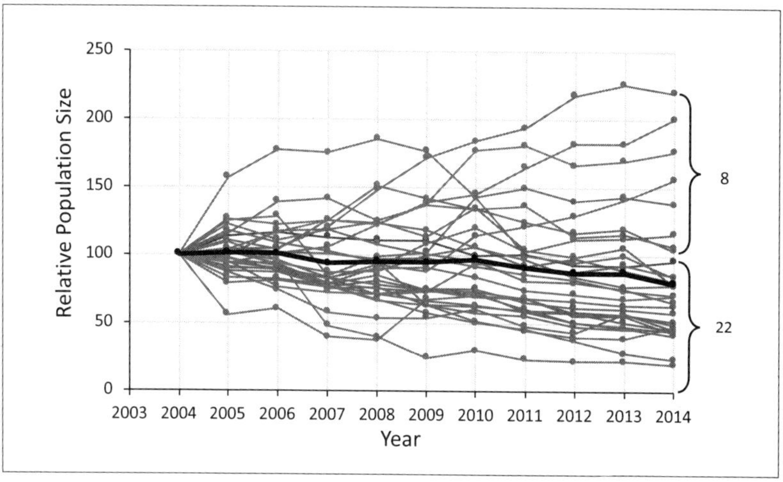

FIGURE 4.6.

Plot showing "optimistic" fate of thirty ginseng populations over ten years, with all starting numbers relativized to one hundred

tions below and see that small size alone does not doom a population, though it does make it more vulnerable. This suggests that while we may well be in a race to save this species from extinction, the plant will not disappear next year. Rather, if we do nothing, populations will slowly continue to dwindle, and the plant will become harder and harder to find. In a few more decades, if nothing changes, it will no doubt be much rarer, but the pattern of decline shown here indicates it will still exist in the wild. Some fraction of populations will be extirpated, but most will be diminished to very small numbers. I don't want to end this discussion with too much optimism, however, because very small populations have their own problems. Problems that can ultimately lead to extirpation.

From the optimistic scenario shown in figure 4.6, we see a pattern very similar to that of the pessimistic scenario, with the same populations doing well and the same populations doing poorly. By including the more optimistic view of seedling recruitment to the population, the average decline was now only 20 percent over the decade (2 percent per year) rather than 30 percent, and now eight populations showed growth over a decade relative to their initial starting sizes. However, again, statistically, more populations declined than increased, and the take-home message is quite similar to what we saw in the pessimistic scenario. All of this suggests a similar long-term fate for ginseng populations—a slow decline for most populations and a long, drawn-out walk toward zero.

Let's take a closer look at what happened in our small populations (fig. 4.7). These populations all began the decade with less than fifty individuals and thus would appear to be the ones most vulnerable to extirpation. Indeed, three of the four declined steadily during that decade. None of these four populations was harvested over that time frame. Therefore, we cannot blame harvest for the decline. The largest of these (Population 11) was located on a flat, northern site with atypical, thin soil. While this population produced large numbers of seeds, seedling recruitment was quite rare, probably due to the challenging soil conditions. Population 20 also declined. This group was subject to high rates of deer browse, and it was found near the bottom of a steep slope. Seeds produced by the adults here likely washed into the stream below the population and died at a high rate, such that, again, recruitment was limited. Population 18 declined the most during the decade. This population was again in what might be considered a marginal habitat, dominated by an overstory consisting of oaks, hickories, and other trees and shrubs suggesting a relatively dry site compared to most ginseng habitats. Compounding the effects of drought here, which reduced new seedling recruitment, was a heavy diameter-limit cut

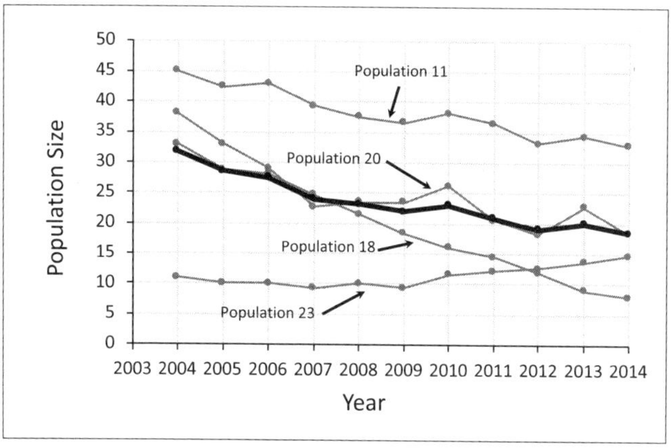

FIGURE 4.7.
Dynamics of four small ginseng populations over ten years. Actual numbers are shown.

of the overstory forest in 2015, after the decade depicted. This was followed by heavy regrowth of a dense, weedy understory, yielding intense competition for the remaining ginseng plants.

The smallest population of the group in 2004 (Population 23) experienced none of these challenges. It had very low browsing, was located in an area with large, older growth overstory and deep soils, and was protected from drought by being located above a stream in a relatively cool ravine. At first, recruitment was limited in this population, but as individual adult plants aged and grew in size, seed numbers, and ultimately new seedling numbers, increased, and the population began to expand.

From the small populations in the study, we understand once again that extirpation does not happen overnight, even when they are in decline. Ginseng is tenacious. However, a population like Population 18, consisting of only eight individuals, is perilously close to gone. Indeed, given its recent past, it may just be a matter of time, without any additional stresses. And a single harvester, digging all eight roots, could put it over the edge.

Attending public meetings on ginseng over the past quarter-century, you would hear many divergent opinions about wild ginseng's fate, accompanied by precious little data. No longer. We have data to guide ginseng conservation now.

Being scientists, however, we were not entirely comfortable with our assessment based on monitoring alone. Certainly, all of our conclusions rested on the assumptions we made in the process of producing figures 4.5–4.7. Most of these assumptions had to do with the fate of seeds. Being small and hard to track, this invisible part of the population presented a challenge.

To get an even better handle on ginseng populations, we needed to study seeds, in addition to the aboveground population, and include them in any predictions of population sizes. This is exactly what we did, as described in the next chapter. Again we will see that ginseng is one of the best species for studying plant population change. Unfortunately, incorporating seeds and making projections into the future will take us into the realm of population modeling. But don't despair—it is not too difficult. Moreover, it is not even necessary to comprehend the model details in order to understand what they are telling us. Onward!

CHAPTER 5

From Monitoring to Demography

> Matrix population models—carefully constructed, correctly analyzed, and properly interpreted—provide a theoretical basis for population management.—**HAL CASWELL**

I recommend choosing one of three ways to read this chapter: (1) read it all, delve deep, and wrestle with the calculations, (2) read it normally, but skip the boxes titled "Calculations," or (3) skip to the last paragraph, grab the take-home message, and move on to chapter 6. I considered relegating this entire chapter to an appendix or even exiling it to a web page, where only the deeply curious among my readers could go to learn how population biologists summarize plant life histories and derive their population growth numbers. You do not absolutely *need* to understand how we do demography in order to read later chapters. If you are someone who is willing to accept scientific conclusions without seeing all the evidence behind them, choose the third option above. This material won't be on the test.

However, if you are curious how ginseng's growth, survival, and reproduction change with age, and if you want to understand why age is a poor predictor of readiness for harvest, and if you would like to understand at least some of the math behind reliable population growth rate estimates, then I invite you to read on here. You will not need math skills beyond basic addition, subtraction, multiplication, and division.

I also decided to forge ahead and present this information in a book aimed at the general public because this is precisely the chapter you would not get in a book written by a science journalist. But, as the cliché goes, the devil is in the details. I prefer not to gloss over all the details. All of our endeavors in science, such as hypothesis formulation and testing, are based on healthy skepticism. Scientists are constantly inviting skeptical review of their proposals and publications from fellow scientists. Why shouldn't we do the same from the public?

What follows is a scientist telling you: "Don't just trust us, find out what we do, why, and critically evaluate the methods we use."

How Does a Ginseng Plant Grow, Survive, and Reproduce as It Ages?

It is safe to say that we humans are unique among species on earth in pausing on the day of our birth to celebrate adding one year to our age. Doing this every year makes us acutely aware of our age and compels us to think about how we have changed and what might lie ahead. We instinctively compare ourselves to others our age, as well as younger and older friends. Awareness of aging, and the fact that there are limits to our age, reinforces the knowledge of our own eventual demise, which in turn has many implications for our thoughts and behaviors. I mention all of this because we might think we generally understand biological aging.

Aging in plants is different. We know that in humans, with our high degree of parental and medical care along with social support systems, our survival rate remains high right up until late middle age, when diseases and physiological breakdown catch up with us. Humans, if they are lucky, can live to a hundred years and perhaps a handful beyond. Then we are toast.

In the natural world, the life history of a human is actually very odd. Wild plant populations, for example, with no medical care and only a modicum of parental care, are vulnerable from the start. The greatest losses occur early in the life cycle, and very few plants reach old age. Think about how many acorns are produced by a huge oak tree, yet on average only one of them will grow up to be another large oak tree! Upon reaching old age, some plants are nearly immortal because they can clone themselves, creating new offshoots and casting off the old parts. All this means we cannot directly transfer our understanding of human aging to plant aging, either at the individual or population level.

In chapter 4, we simply added births and subtracted deaths to observe the change in a population from one year to the next. This is fine for monitoring. But population biologists realized early on that if we wanted to *predict* future population change and not just look backward at what had happened, monitoring was not enough. Most species, plants included, have birth and death rates that depend on how old they are, at a minimum. Thus, you would think everyone interested in the future of a species would want to know how birth rates and death rates change with age. Although this was the bread and butter of animal population biology, there have been remarkably very few sustained efforts to carefully quantify age-specific birth rates and death rates in plants. It has

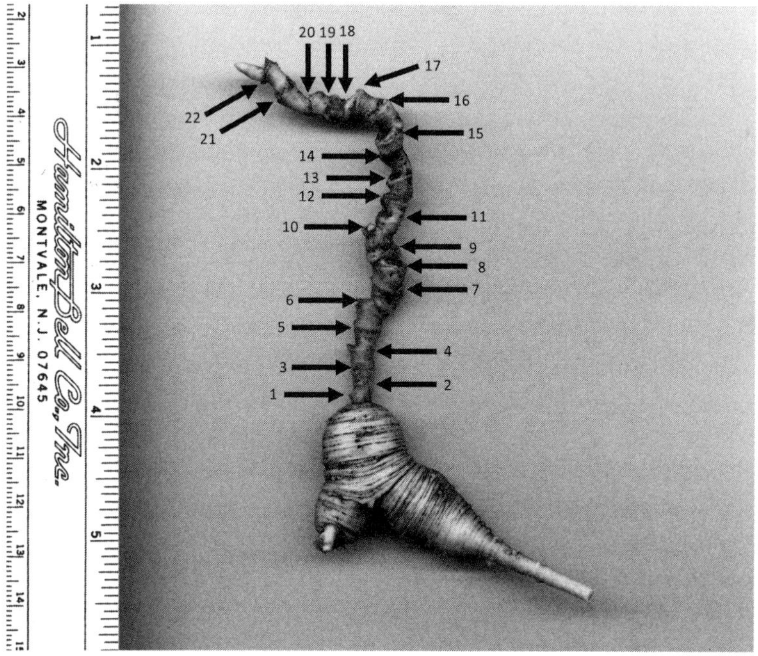

FIGURE 5.1.
Image of a twenty-two-year-old ginseng root. Each bud scar along the rhizome (neck) marks one year of aboveground growth, so counting the scars yields age of the plant. (Photo by Eric Burkhart)

been done for short-lived plants such as annuals or biennials, and occasionally for trees where annual growth rings make aging easier. However, that leaves a big gap in our understanding of perennial plants that do not have annual rings.

Fortunately, our twenty years of following populations in the field gave us important insights into the age-specific births and deaths of ginseng plants. These insights are all the more important because current CITES regulations limit harvest to five-year-old or older plants, and this is virtually the only verifiable harvest regulation in place. Only with knowledge of birth and death rates with age can we tell whether this limit on wild harvest is effective at preserving wild ginseng.

How can we determine age-specific births and deaths for a plant that dies back belowground each winter and does not have growth rings? Luckily, we have two tools for doing that with ginseng. First, we can follow new seedlings (age zero), then add one to their age in our database on their birthdays each year thereafter. By tagging and censusing groups of these seedlings (also known as cohorts), we can mathematically determine the proportion dying at each age and, if the individual does survive, the average number of seeds these survivors produce at each age. This first approach could be used with any plant, but this method would limit us to the number of years we followed each marked plant. However, ginseng can and occasionally does live twenty, thirty, forty, or more years.

Fortunately for us, older ginseng plants can be aged a second way—by counting bud scars on the belowground stem, or rhizome (fig. 5.1). Aging plants by counting rhizome bud scars is not flawless, mainly because some of the early scars are very small and hard to see, and when the plant is rooted in place it is difficult to see all of them. Nevertheless, we can get an approximate age of much older plants with this approach. From our thirty ginseng populations, we selected a subset of twelve on which to perform rhizome aging, ultimately obtaining a sample of 683 plants. Of course, every year thereafter, we knew their ages as well by adding one to their age each spring.

Once we had a set of aged individuals in the population, we summarized the data in a table that simply shows how births and survival changed with age (table 5.1).

CALCULATIONS

You will notice in the table that survival and birth can be put in the same terms. Both rates are quantifying the number of individuals going *to* a particular age class (the row) *from* another age class (the column). Survival is shown in the stair-step of numbers from upper left to lower right in the table. In the case of survival, this will always be a fraction ranging from 0 to 1. In other words, it gives the number making it to May of year two divided by the number present in May of year one. If we multiplied those proportions by 100, this would be "percent survival," so it's easy to see what those numbers mean. For example, the percent survival from age 0 (these are new seedlings) to age 1 is 69 percent (.69 × 100).

Births can also be thought of as number of individuals going *to* the seed class *from* each adult of a given age. Across the top row of the table, I show average seed numbers for this population, which is why these "fertility" numbers are so small. For example, the average seed numbers in the soil in May of year two per adult of age seven in May of year one is 0.125. Individuals that produced seeds were averaged in with many that produced none. This particular population had very low seed production, but in theory these numbers could be greater than 1.0. In other words, the average seed production per adult of a given age could be more than 1.0. But ginseng is generally a low seed producer under natural conditions.

TABLE 5.1.

Birth rates and death rates over a one-year period, Population 27

Shown are actual numbers for Population 27, which was censused for twenty seasons from 1998 to 2017, with births in top row (mean seed numbers from plants age zero to ten) and survival chances in the subdiagonal line (proportion of adults successfully transitioning from each age to the next age). The data have been pooled across all years to obtain an adequate sample size. Note: the census each year was done in May or early June when new seedlings had just emerged and were zero years old. The nonblank cells in the table represent change from one spring to the next spring. This table is only a partial one; not shown are ages beyond ten years.

			FROM											
			Age Year 1											
TO	Age Year 2		Seeds	0	1	2	3	4	5	6	7	8	9	10
		Seeds	?	0	0	.016	0	.040	0	.024	.125	0	0	0
		0	.76											
		1		.69										
		2			.80									
		3				.79								
		4					.86							
		5						.84						
		6							.83					
		7								.90				
		8									.90			
		9										1		
		10											.88	

TABLE 5.2.

Birth rates and death rates over a one-year period, Population 30

Shown are actual numbers for Population 30, which was censused from 2001 to 2017.

The data have been pooled across all years to obtain an adequate sample size.

			FROM											
			Age Year 1											
			Seeds	0	1	2	3	4	5	6	7	8	9	10
TO	**Age Year 2**	Seeds	?	0	0	.03	.16	.31	.08	.11	.57	.83	.63	0
		0	.76											
		1		.86										
		2			.91									
		3				.90								
		4					.95							
		5						.97						
		6							.89					
		7								.95				
		8									1			
		9										1		
		10											1	

By summarizing the data this way, we can quickly see patterns. First, the survival rates shown in table 5.1 are remarkably high for a plant. More than two-thirds of new seedlings survive for a year or more. This suggests that these seeds are very well provisioned by the mother plant (maternal care!) and that the environment is benign. By age seven, survival is higher than 90 percent. Second, we also see that these plants are not replacing themselves quickly. If we add up the fertilities, by age ten the cumulative fertility is only around 0.2 seeds. Clearly this is not enough for a plant to replace itself by age ten. And this does not even account for the fact that many plants won't survive to that age.

Just how variable are age-specific birth and death rates in populations? An easy way to look at this is to compare Population 27, summarized in table 5.1, to Population 30, which is doing very well (table 5.2). At first the differences do not seem dramatic. Survival rates were very similar, although you can see that on average they were a little higher in Population 30. The main difference you can see is the consistently higher seed production rates.

When I show these numbers to ginseng growers, they are almost always very surprised, and some even think I must have made an error. When ginseng is cultivated, seed production can be much higher. If you are a gardener, however, you may not be at all surprised. Plants are very plastic, meaning their growth and reproduction depend greatly on the conditions in which they are grown. In the case of natural populations, no one is tilling, weeding, fertilizing, or spraying insecticide or fungicide on these plants, and the first thing to drop is growth. Since seed output is less in small plants, very low seed production for the first ten years of life is the result. If you forget to weed your garden, the result will be the same!

The Seed Bank

Careful readers will note that tables 5.1 and 5.2 are incomplete. For one thing, they only show the first ten years of data. But also there is a "?" in the table representing the transition from seed to seed. What is this missing information? As discussed previously, ginseng has a population of viable seeds sitting in the soil at all times. When a red berry drops from the plant, usually in September or October, the seed within does not germinate immediately. Indeed, with ginseng, that seed is not even able to germinate the next spring. Instead, the tiny embryo inside the seed must develop through at least one more summer before germination can occur the following spring. Some seeds may take two additional summers, and a few even take three summers prior to germination.

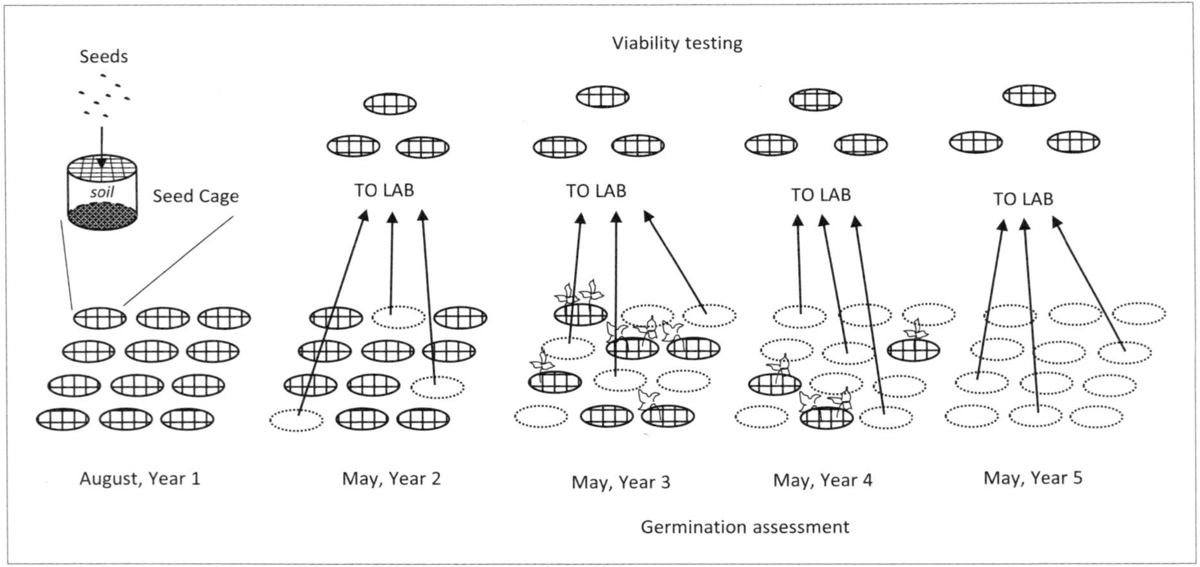

FIGURE 5.2.
Seed cage study to track seed fates in natural field soils

TABLE 5.3.
Seed bank dynamics from spring of year one to spring of the following year

			FROM			
			Seed Age Class (year 1)			
			9 mo.	21 mo.	33 mo.	45 mo.
TO	Seed Age Class (year 2)	9 mo.				
		21 mo.	.07			
		33 mo.		.20		
		45 mo.			0	
	New Seedlings		.82	.25	1	

We uncovered the age-dependent dynamics of the "seed bank" in ginseng by performing a controlled experiment (fig. 5.2). We created seed cages consisting of short sections of PVC pipe with window screen glued to the bottom. The cages were filled with soil from the site, and then berries were added containing a total of fifty ginseng seeds. In order to obtain adequate seed numbers for the study, seeds were obtained from ginseng forest farmers who "lightly" cultivated plants under a forest canopy. Finally, we added a plastic hardware cloth top to prevent berries from being washed away, and seed cages were set down into a hole in the soil at the field site where the natural population was found. In other words, we tried as hard as we could to replicate the microenvironment of a seed in the soil of that site.

Twelve replicate seed cages were placed at each of the thirty natural ginseng populations for a total of 360 cages containing a total of eighteen thousand seeds. Each spring thereafter we counted germinating seedlings in each cage. We also took three cages back to the lab each spring. The removed cages were sieved to extract the seeds. The final step was to slice the seeds open and expose them to a dye that turns the flesh pink if the seeds are still alive. With this elaborate and time-consuming procedure, we calculated the germination and survival rates of seeds as a function of their age in the seed bank, with separate numbers for each of our field sites.

The fact that ginseng's natural seed bank in May or June contains seeds that may be nine, twenty-one, thirty-three, or forty-five months old complicates our neat life history table (tables 5.1 and 5.2), but it's all still just a matter of accounting. This means that rather than a life history with one stage and a single from–to cell in our table, our actual life history has two sets of age-specific rates (for seeds and plants). The upper left corner of our from–to table is not just a single placeholder in the table, but something that describes the seed bank dynamics (table 5.3).

Total germination from the seed bank after four years varied between about 40 percent and 85 percent. This range can certainly make a difference to population growth. However, model simulations showed that differences among populations in how much germination occurs early (after twenty-one months) versus late (thirty-three or forty-five months) had very little effect.

When Has Ginseng Replaced Itself?

Now that we understand the age-specific survival and birth rates at both the seed and plant stage, we can proceed to answer interesting questions about ginseng's life history. If we are going to use plant age to determine a cutoff point for harvest, then one possible criterion for the cutoff age would be that age which, on average, ensures that an individual has replaced itself in the population. By the way, this does not ensure that the harvest will be harmless, because removing that individual would preclude future seed contributions for that plant. But if seeds found on the harvested plant are optimally sown, the damage should be reduced by harvesting at that age, rather than earlier.

CALCULATIONS

To answer the replacement question, imagine a cohort of one hundred age-zero seedlings all born in the same year. Then our question becomes how many years would it take for these one hundred seedlings to mature and produce a total of one hundred offspring plants? You can easily envision how we could answer this question, knowing age-specific survival and reproduction. First, we would multiply 100 times the survival rate of age-zero plants to get the number expected at the age of one. Let's do it for Population 1 (table 5.1): 100 × .69 = 69. We know age-zero plants do not produce seeds, so we don't have to account for reproduction yet. Then we multiply the number that survived one year (69) by the survival rate of one-year-olds: 69 × 8 = 55.2. Again, we don't have to account for reproduction because one-year-olds produce no seeds. Then we multiply the number that survived to year two by survival rate to get the number at year three (55.2 × .79 = 43.6). However, each two-year-old plant does produce on average 0.16 seeds, so 55.2 × 0.16 seeds enter the seed bank—8.8. Note that we have not replaced a single new seedling yet because no seeds produced by this group have germinated. Nevertheless, using this tedious process of adding up all new seedlings eventually produced by the original group of one hundred, we could determine the age at which those initial hundred plants had produced one hundred offspring. This kind of accounting is what computers were designed to do. They are far more

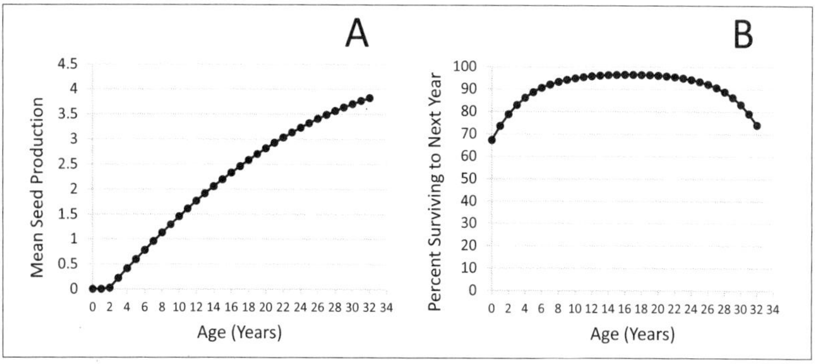

FIGURE 5.3.
(A) Mean fertility measured in terms of seed production, and (B) survival chance, as a function of plant age (data pooled from all thirty censused populations)

accurate than humans, and they are very fast. So we will let them do the accounting and figure out the answer to our question: By what age has an individual new seedling replaced itself, on average?

The answer is shown in figure 5.4, but a little background is needed. First, for the two populations in tables 5.1 and 5.2, we did not have survival or fertility statistics for older plants because we did not age any plants using the rhizome bud scar method at those populations. Therefore, we pooled data from all the aged plants from all thirty populations using both the new seedling tracking method, plus all the older rhizome-aged individuals from the subset of twelve of these populations with rhizome-based ages. In total we had 19,259 "observations" of plant fates for those plants where we had an age. Some of these were repeat observations of the same individual as it aged through time, so they were not necessarily "independent" observations. Nevertheless, this allowed us to determine age-specific survival and seed production with the largest possible sample size. Ginseng enthusiasts are always curious about the vital statistics of ginseng with age, so before addressing the replacement question, I'll show the raw survival and fertility data as a function of age (fig. 5.3). Three features of these curves stand out to the plant population biologist. First, as described before, ginseng has very low seed production in natural populations (fig. 5.3A). Each seed produced is fairly large, given the size of the plant, so the trade-off seems to be that even older, larger plants do not produce very many, on average. It is important to realize these are averages, however, calculated from many zeroes and occasional plants that produce numbers far above the average. The

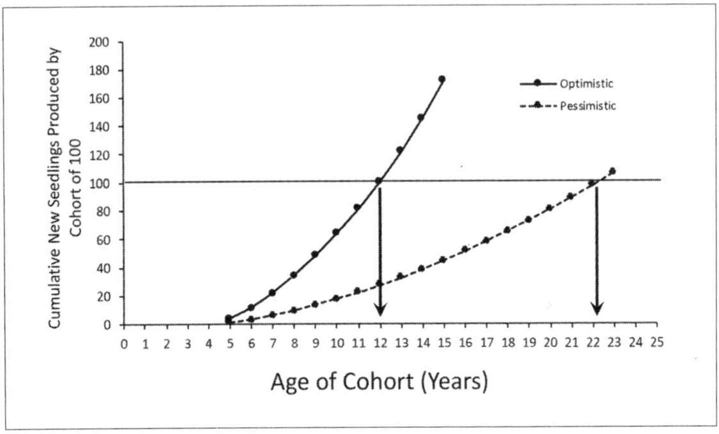

FIGURE 5.4.
Cumulative number of next generation new seedlings produced by an initial cohort of one hundred new seedlings. Numbers used survival and birth rates pooled across all thirty ginseng populations. Arrows show the age of the cohort when one hundred next generation seedlings have been contributed to the population, signaling replacement of the original one hundred.

second outstanding feature is the high survival rate (fig. 5.3B). As discussed before, the fact that two-thirds of new seedlings will likely survive their first year is amazing. The survival rate of adults between age ten and twenty, 95 percent, is also startlingly high, especially given that this is a harvested plant. Indeed, the death rate of these plants (5 percent annually) is not out of line with the harvest rate in natural populations. The third feature that stands out is the falloff in survival beyond about age twenty-five. The reasons for this are unclear, but a form of aging may be responsible. Certainly trees do decline beyond some age limits in most environments, so this should not be too surprising, but we have a very poor understanding of aging in natural plant populations generally. Are they accumulating bad mutations? Are they acquiring diseases that eventually kill them in the end? We don't know. What is the oldest ginseng plant ever found?

Back to our replacement question. Given the survival and seed production curves shown in figure 5.3, how long does it take for one hundred new seedlings to contribute one hundred offspring to the population (i.e., to replace themselves)? Figure 5.4 shows two answers to that question: twelve years ("optimistic") and twenty-three years ("pessimistic"). Most likely the actual answer lies somewhere in between.

The optimistic scenario assumed that seeds survive at the high rate seen in

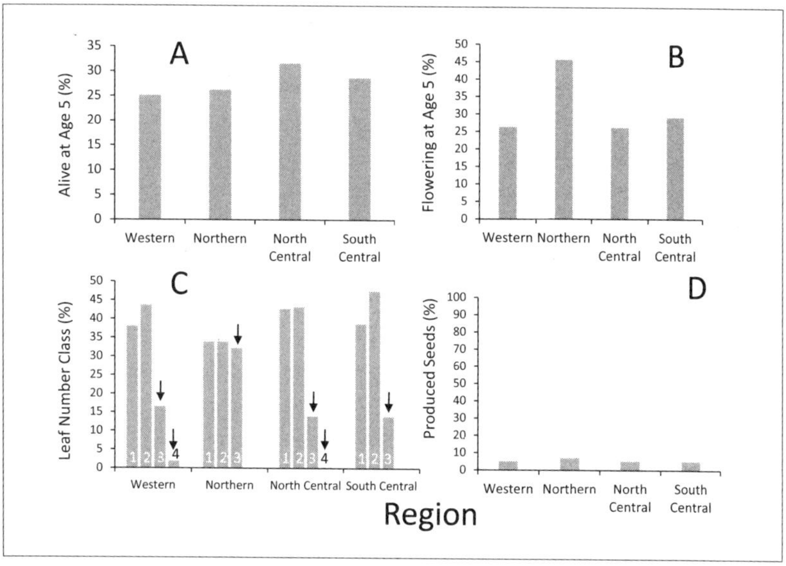

FIGURE 5.5.
Statistics on five-year-old individuals in thirty natural populations of ginseng separated by region. (A) Net survival from germination (age zero) to age five for ginseng seedlings, (B) flowering, (C) size distribution by leaf number class, and (D) percent that have produced seeds by age five.

the seed cage experiments. Since seed cages had rims that would prevent washout of seeds, and plastic hardware cloth covers that would prevent some predation, we suspect survival in the seed cages was higher than that for naturally dispersed seeds.

The pessimistic scenario adjusts the germination rate to reflect what we actually observed in the field. In our censuses, we routinely scanned for new seedlings within two meters of any existing plant in the census. Earlier studies suggested that 90 percent of seeds stayed within this perimeter, so we assumed when we recorded new seedlings that we were recording 90 percent of those that were actually produced. This pessimistic estimate did not adjust for bird dispersal of seeds outside the two-meter ring around each plant we censused.

The mean of the optimistic and pessimistic scenarios is 17.5 years. Again, to put this in perspective, it is important to realize this number is from a pooled data set for all thirty populations. But as we have seen, the age-specific rates differ among populations, so in fact the replacement age would be lower in fast-growing, healthy populations, and higher in slow-growing populations in suboptimal environments.

Should a harvester feel good about harvesting only those plants that have replaced themselves? It depends. If the population is declining, by harvesting only those plants that have replaced themselves (i.e., those that are doing better than average in the population), virtually *any* harvest is actually hastening the decline. But certainly using this average criterion would be better than the current age cutoff of five years.

Characteristics of Five-Year-Olds

What we *can* definitely conclude from this analysis is that the five-year age limit for harvest is clearly not sufficient to ensure that individuals have replaced themselves. This becomes even more obvious if we examine the statistics for five-year-old ginseng plants more closely. I compiled data for all new seedlings for which we had five years of data following the year of germination and summarized their survival, reproductive, and size traits. From these data, we first see that about one-fourth of seedlings survive to year five (fig. 5.5A). Again, these are remarkably high numbers for a plant, reflecting the proficiency with which seedlings extend a storage root that can ensure survival from one season to the next. Nevertheless, the first five years do represent a vulnerable phase when this root is just getting established. Later in life, the root reliably provides energy for regrowth in the spring and access to water and nutrients to survive stresses such as deer browse or drought. There was a slight tendency for plants in the center of the range to survive at higher rate than those in northern or western populations, but the difference was small.

The five-year rule as a threshold for harvest is intended to ensure reproduction prior to harvest, thereby limiting damage. How well does it work? At age five, the majority of plants in all regions are not flowering and therefore are incapable of producing seeds (fig. 5.5B). Clearly, in natural populations, these plants are immature. Almost all three-leaf plants produce flowers every year, so this implies plants are mostly smaller than this. The one significant regional difference in flowering rate was the higher amount seen for northern populations (in this case, from Pennsylvania and New York), suggesting these plants grew somewhat faster. Even there, however, less than half of the plants were flowering (fig. 5.5B).

Figure 5.5C shows the size structure of five-year-old plants in populations grouped by region. The majority of plants have not reached "adult" size by age five. In fact, in all four regions, more five-year-olds were still in the one-leaf stage than the three-leaf stage, reflecting the slow maturation of these plants. All states require ginseng to have three or more leaves before they can be harvested; however, since leaves are not turned in by diggers along with the roots, no verification can be done at the point of sale. A ginseng dealer buying a batch of five-year-old roots would have no reason to refuse the sale based on the evidence in front of them, but figure 5.5C shows that a large portion of their purchase would be illegally harvested roots. This is a clear point of breakdown between the intent of the regulations and the enforceability of them.

Perhaps the most convincing evidence that the five-year rule does little to ensure sustainability is the fact that only about 5 percent of new seedlings produce a single seed by the age of five (fig. 5.5D). Since we already examined the "replacement age" and found it to be between twelve and twenty-three years, perhaps this low seed output at age five should not surprise us either, but it does show how far off the five-year rule is in terms of affording *any* protection to the species.

You might reasonably object that an ethical harvester would also follow state rules specifying that plants must have three leaves before they can be harvested. However, one-third of five-year-old plants *with three leaves* have never produced a seed. The average lifetime seed production of age-five, three-leaf plants is 3.3, which is more encouraging, but not nearly sufficient for replacement, given losses in the seed bank and mortality between the new seedling stage and age five. Harvesters can reduce losses in the seed bank by carefully planting seeds at an optimal depth of one-half to one inch, but even if none were lost prior to germination, at least four new seedlings would need to germinate to ensure that one of them would reach age five. And recall, most of these would not yet have three leaves.

We humans base many policies on age because our own brains and bodies develop over time in a predictable manner. School attendance, driving, voting, drinking alcohol, military service, retirement, and many more policies are tied to age precisely because age is an indicator of competency. We might naturally assume other organisms follow an age-related program as well. But for plants, survival, growth, and reproduction are much more closely related to stage or size than to age. This would not be a problem if age was tightly related to size, but a simple plot of age versus size (leaf area) for our ginseng plants shows that this is not the case (fig. 5.6).

Some of the scatter we see in figure 5.6 is due to site differences, but most of it is due to differences between individuals within populations. Some are in good micro-sites in the population, and some are in poor ones. Some have been browsed by deer, while others have not. Some have been eaten heavily by insects, while others have not. Some are in pockets of richer soil, and others are not. Some may have inherited "good genes" for the environment they are in, while others are poorly adapted. For a wide variety of reasons, ginseng may

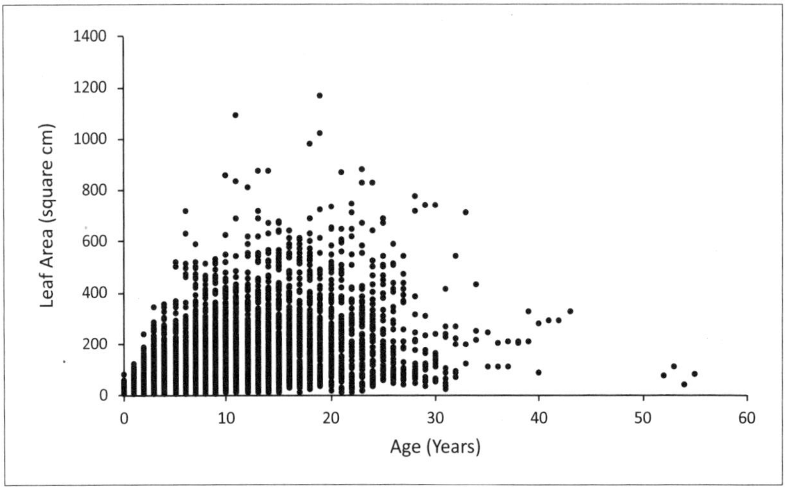

FIGURE 5.6.
Relationship of age to leaf area for 19,230 measurements

survive and grow at very different rates. Since size is a good predictor of survival and reproduction, while age is only loosely related, this suggests that we might be better off developing a new way of examining population growth, using size instead of age. In the next section of this chapter, that's exactly what we'll do.

Oh, and the number is 132—132 years old, that is. Reportedly the age of oldest root of American ginseng aged by the bud scar technique. So ginseng plants can outlive humans. But they rarely do. Something usually gets them, whether it's disease, harvest, drought, or repeated deer browse. The life of a ginseng plant is hard, and to live that long is to be extremely lucky.

Predicting Future Population Growth: Size Matters

Predicting the future of natural ecosystems is a lot like forecasting the weather. Based on current conditions and our current understanding of factors that govern the system, we can make projections of fate. But virtually always these projections involve uncertainties because, in the real world, forces we do not understand or anticipate enter in and alter the course of events. In weather forecasting, the magnitude of these uncertainties is somewhat known, so meteorologists can forecast a particular chance of rain three days from the present. Their forecasts get more and more accurate as the time horizon gets shorter—a forecast for next month is far less accurate than one for next week, which in turn is far less accurate than the forecast for tomorrow. This is because the number of unknown factors gets smaller as the present approaches the forecasted time.

In the same way, when ecologists predict a certain population change, it is important to view this like a weather forecast. Treat any very long-term prediction with a great deal of skepticism, but shorter-term predictions are likely to be more accurate.

Another way to view population forecasts is to think of them as "if-then" experiments. Specifically, if conditions in the future remain as they were when birth rates and death rates were measured, then the predicted outcome *will occur*. No one expects conditions *will* remain exactly the same as during the measurement period, but, to the extent that they do, the population projections will be accurate. In other words, the mathematics are right, but the assumption of constant rates underlying a specific forecast may or may not continue to hold. You can assess the consequences of births and deaths being larger or smaller than they are at present, and you can speculate about what could occur in the future to make them so. Indeed, with a long measurement period, we can quantify the normal variation in births and deaths that occurs and place upper and lower bounds on population growth rates.

This long-winded and somewhat philosophical introduction to forecasting or prediction is needed so you understand how ecologists and conservation biologists think about theoretical models. If you think of predicting the future of ginseng as a useful modeling exercise—one that explores what *might* happen—then you are in a great frame of mind to interpret the results I am about to describe.

Ginseng Life Cycle

As we saw earlier in this chapter, individuals in a population have chances of dying and chances of giving birth that change with age. For example, we know a newly germinated ginseng seedling has zero chance of producing a seed (they don't even produce flowers in the first year) but a fairly good chance of dying (that first year of life can be pretty stressful for a little plant). We also saw, however, that age is a pretty poor predictor of survival and reproduction and that the "stage" of the plant is often quite a bit better. For example, if a ginseng plant is five years old and is still a one-leaf "seedling," then its chance of producing a

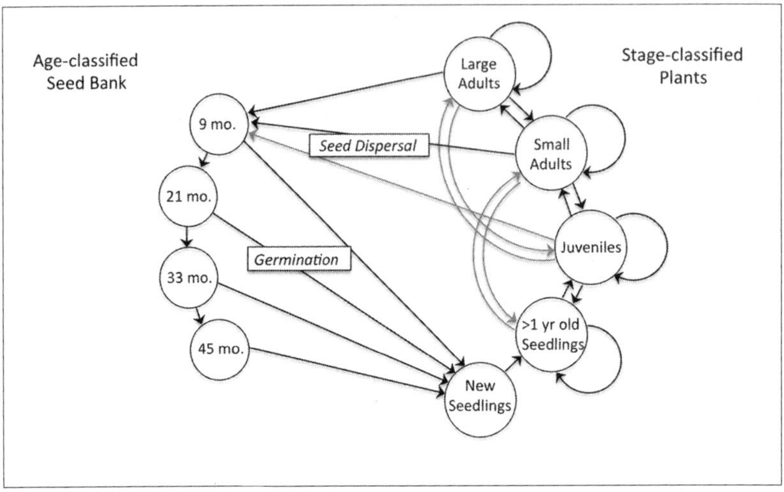

FIGURE 5.7.
Life-cycle diagram of American ginseng showing nine life history stages (within each circle) and possible transitions (arrows) among them from one year to the next. Lighter arrows show transitions that are possible but occur at lower rates. The "state" of the population (i.e., the number in age or stage) is assessed near the beginning of the growing season (end of May, early June).

seed is still zero. But if a ginseng plant has germinated in a fertile site, and it has been lucky enough *not* to have been browsed by a deer or damaged by some other factor, it could have grown to the three-leaf adult size, in which case the chance of producing one or more seeds is very good (about two out of three). For this reason, models of ginseng and other plants typically determine birth and death rates as a function of stage or size rather than age.

Figure 5.7 shows a typical life-cycle diagram of ginseng. Different population biologists have used alternative versions of this depiction, but all of them are very similar. The very first descriptions did not take account of the "seed bank" that is stratified by age because the seed cage experiments had not been done yet. Different formulations of the aboveground stages have been used as well; sometimes new seedlings and seedlings more than one year old are lumped into one class. Although all of the life-cycle diagrams have defined adults as plants with three or more leaves, small and large adults may be defined differently, using leaf number or leaf area, or perhaps dividing adults into more than two classes. Juveniles are defined as two-leaf plants because these have been viewed as transitional between seedlings and adults, but occasionally they produce seeds successfully.

CALCULATIONS

Population biologists turn the life-cycle diagram into a mathematical model by attaching numbers to each of the arrows. For example, the arrow from large adults to the nine-month-old seed class will be given a number equal to the average number of seeds produced by individuals in the "large adults" class.* The same is true for all the other arrows going into the nine-month seed category. Arrows in the diagram not leading to the nine-month-old seed class will be numbers between zero and one, giving the number of individuals for the stage at the end of the arrow and per individual in the stage at the beginning of the arrow. For example, for the arrow from nine-month-old seeds to twenty-one-month-old seeds, this would be the chance that a nine-month-old seed survives to become a dormant, viable twenty-one-month-old seed the next year. Or the arrow from the "new seedlings" class to the class of seedlings over one year old is simply the chance that a new seedling will survive to the next year, because that is the only transition that it can make, other than dying. Most of the other arrows are not strictly survival rates because there are multiple possible fates. For example, the arrow from the "juveniles" class to the "small adults" class is the chance that a juvenile will become a small adult, but juveniles can also survive and remain juveniles (the looped arrow back to the juvenile class) or regress to the smaller class of seedlings over one year old, or, though rarely, jump to the "large adults" class. All of the quantities associated with arrows in the diagram are "transition probabilities," using the parlance of the trade.

*Since transitions are from one spring to the next, arrows leading to the nine-month-old seed category are not quite this simple; in fact, we need to multiply seed production, which occurs in August or September, by the survival rate of those seeds over the following nine months. We determined this survival rate using the seed cage studies described above.

For population biologists, the purpose of the long and arduous process of censusing populations in two consecutive years is to fill in the numbers in the life-cycle diagram. These numbers are interesting in themselves, especially for making comparisons. For example, the numbers may differ among years, and when we have many years of data we can begin to ask what factors are controlling the numbers. Were many plants browsed by deer that year, lowering seed production? Was it a hot and dry year, so survival was lower? Was there more disease in the population that year? We can also look for differences among populations and again, with many years of data, begin to decipher why those differences exist.

A second purpose of the life-cycle diagram is to construct a predictive model of population growth. As promised, you will be spared the mathematical details. If you are truly interested, you could look at Hal Caswell's famous book on the subject, *Matrix Population Models*. That book will give you more than you want to know. Suffice it to say, in population ecology, there have been hundreds, if not thousands, of such models formed for "wild" species of plants and animals. Almost always, these models suffer from not enough field data in one way or another. We are incredibly fortunate with ginseng to have both an easy plant to work with (see chapter 3) and a huge data set that covers a large number of years over a wide range of populations.

What can we learn from these models that we cannot get from the monitoring described in chapter 4? First, we can forecast future population numbers over the short term. Monitoring just shows what happened, not what *will* happen. Second, we can project the population growth rate over the long term. Finally, we can tease apart which arrows in the life-cycle diagram are most important as controls over the population growth rate. If you wade into Caswell's book, you will find *much* more that can be done with these hard-won "transition probabilities," but for now we will stop there. For the rest of this chapter, I will focus mainly on the projected long-term population growth rate.

Long-Term Population Growth Rate

How do population ecologists quantify "long-term population growth rate"? While again avoiding complex mathematical derivations, we do need to wrestle with this concept briefly. When we make forecasts for a population with stage classes like those described in the life-cycle diagram (figs. 5.7 or 5.8), the fraction of individuals in each class bounces around a bit in the first several years of the forecast. Of course, the sum of those fractions has to add up to one, but the fraction in each class will generally not remain the same from one year to the next. This happens in human populations too. For example, in the United States during World War II, men leaving in droves for the battlefield naturally resulted in a very low birth rate back home. But when they returned home after the war was over, the birth rate soared. This produced the generation known as baby boomers. So the fraction in the age classes containing baby boomers was unnaturally high. There was even an echo of the baby boomers when they began to have children, but it was dampened slightly because they had children at a variety of ages. As baby boomers have become grandparents, the echo of the echo has become even more muted.

Likewise with plants. In theory, without a major disruption, the fraction of individuals in each class eventually begins to stabilize over time. When that happens, the year-to-year population growth rate also converges toward a single value called "the stable population growth rate." Now "stability" in this context does not mean zero growth; instead, it means a constant rate of growth.

CALCULATIONS

You can think of the stable population growth rate as a multiplier. It is the number you would multiply the current total population size by in order to get next year's population size, after the fraction in each class has stabilized. For example, if a population was growing at a stable rate of 12 percent per year, the stable population growth rate would be 1.12. In other words, 1.12 is the multiplier that would be needed to show that 12 percent increase. You can easily see this by putting in a population size of one hundred and doing the multiplication (100 × 1.12 = 112). Easy, right?

The same is true for decline. For example, if a population was declining by a stable rate of 12 percent per year, then the stable population growth rate would be 0.88 (100 × .88 = 88, a 12 percent decline).

From the above description, then, a population that is unchanging would have a stable population growth rate of 1.0 (100 × 1 = 100, or 0 percent change).

If you understand population growth rates as varying around 1.0, with values larger than 1.0 indicating expansion of the population, and values less than 1.0 indicating a declining population, then you are ready to proceed to the findings for American ginseng. Clearly, if we are concerned with conservation of this species, we want to know whether the population size is level or growing in wild populations, or whether it is in fact declining.

Ginseng Population Growth Rate: Early Models

Two Canadian scientists, Danielle Charron and Daniel Gagnon (1991) performed the first serious ginseng population modeling efforts using the easy metric of leaf number as the way to categorize individuals. They studied four natural populations at the northern edge of the range in Quebec. Although the combined number of plants was less than four hundred, and the study included only two growth transitions (over three seasons), their initial basic insights have largely stood the test of time. They established that ginseng had a "slow" life history, with a long pre-reproductive period, slow individual growth rate, and high survival of established individuals. The stable population growth rate varied around 1.0, with values as low as 0.87 and as high as 1.19. Given the limited duration and sample sizes of this study, generalizations about overall growth or decline were not possible. Although Charron and Gagnon performed a crude estimate of harvest potential, their results varied widely (from 0 to 15.8 percent could be harvested "sustainably" by this estimate), but the assumptions about harvest used other authors' methods and were not very realistic when applied to ginseng.

Several other studies nevertheless recognized the power of the size-based population models and used them to examine harvest, deer browse, climate change, and forest canopy disturbance with larger sample sizes. We will examine each of those in upcoming chapters on those topics. For now, I want to instead exploit the full power of the long-term data set we began assembling in 1998, continuing through the summer of 2016. I will focus on the decade from 2004 to 2014 within this period because we had complete data for all thirty populations over that time.

Ginseng Population Growth Rate: Thirty-Population Model

The great advantage of a large thirty-population model is sample size. Not only were from four thousand to six thousand individuals measured every year, but also they were repeatedly measured over ten or more years. This allows us to construct perhaps the largest demographic model to date for a wild plant population.

We can also fine-tune the size classes so they capture more subtle differences in survival, growth, and reproduction. The best way to do this is to increase the number of size classes, narrowing the width of each one; however, this is only possible when the sample size is adequate to calculate all the added "transition probabilities" from one year to the next. The life-cycle diagram can be redrawn to reflect this refinement (fig. 5.8).

For a very big data set such as ours, we could have almost fifty classes and still get very good estimates of the numbers attached to each arrow (fig. 5.8). We now examine the distribution of individuals among size classes, the patterns of survival, growth, and reproduction as a function of these classes, and finally the population growth rates projected by these patterns.

FIGURE 5.8.
Ginseng life-cycle diagram, showing forty-nine size-based "stages." Only a partial set of transition arrows is shown in order to avoid a very messy figure. In a complete figure, each stage would have arrows entering and exiting, but the figure shows all of the types of transitions that are possible (growth, staying the same, shrinkage, seed production, germination, and survival). Seed stages are shown with dotted circles. Size classes of plants are based on leaf area. Drawings show one-, two-, three- and four-leaf plants that roughly correspond to the size classes on the outer ring. Each possible transition represented by an arrow has a number associated with it.

The size class distribution (fig. 5.9) is not particularly meaningful in the absence of further information, but we do see two peaks, one around Size Class 15 and one around Size Class 31. These correspond roughly to the cohorts of one-leaf plants (or "seedlings") and three-leaf plants that persist for many years in a population. Extremely small plants (Classes 5–10) are often "injured" or damaged one-leaf plants that are barely hanging on. At the opposite end, the rare huge plants (Class 45 and above) are often the result of some short-term stimulus of additional light reaching the forest floor (e.g., through tree fall).

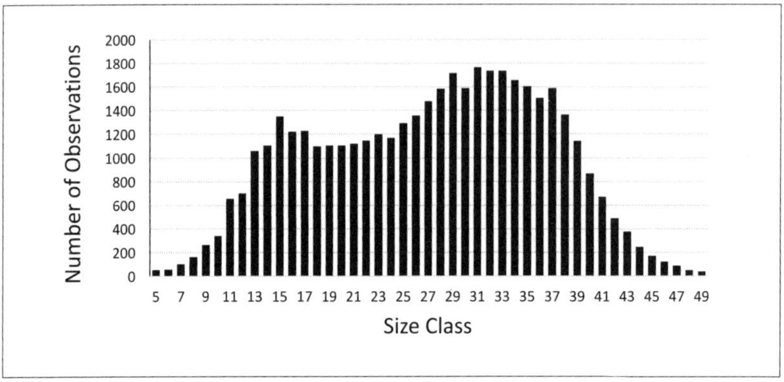

FIGURE 5.9.
Postgermination size class distribution of 42,431 observations of ginseng plants. Size class boundaries are based on leaf area.

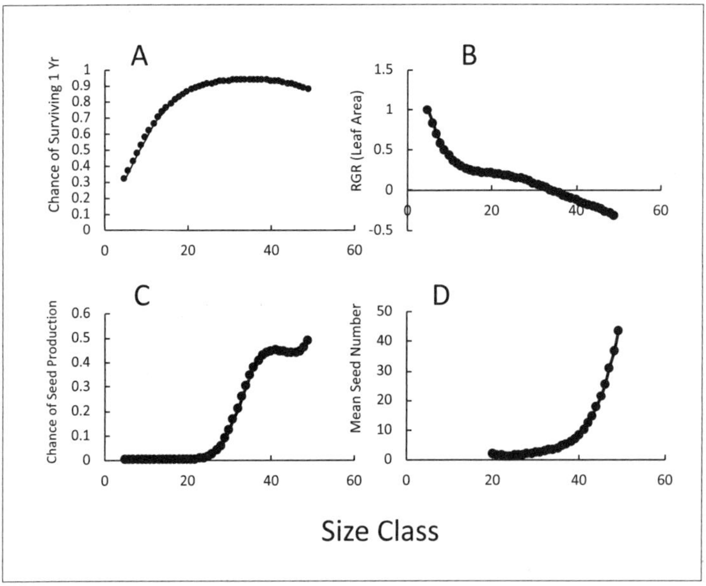

FIGURE 5.10.
Size class versus (A) survival rate, (B) relative growth rate (based on leaf area), (C) chance of producing seeds, and (D) mean seed number (for seed producing plants)

More interesting are the patterns of "life history" traits as they vary with size class (fig. 5.10). Between Size Classes 5 and 20, survival increases dramatically from below 40 percent to more than 95 percent (fig. 5.10A). As they acquire greater leaf area, corresponding increases in belowground root size and depth allow plants to survive all kinds of stresses better, from drought stress in midsummer to the stress of losing leaf area to insect or deer herbivory. Growth, in relative terms, is highest when plants are small, reflecting the importance of escaping the smallest size classes (fig. 5.10B). Then growth rate is low between Classes 10 and 20 but positive as plants mature into adults that increase in leaf area and also begin to allocate energy to reproduction. Growth rate then declines again from Size Classes 25 to 50, actually becoming negative at Size Class 35. This drop in growth rate is happening while the chances of seed production rise rapidly (fig. 5.10C). This is most likely not a mere coincidence. Seed production is costly in terms of energy and resources the plant has accumulated; the "cost" is partly manifested as slower growth. Even plants make sacrifices for their offspring!

The chance of producing seeds rises exponentially past Size Class 25 (fig. 5.10C), corresponding to the greater number of flowers produced by large plants but eventually levels off at about Size Class 40, when evidently some other factor limits whether seeds are produced or not. Could this be availability of pollinators or nearby neighboring plants that are producing pollen or both? We don't know. The last panel (fig. 5.10D) shows that *if* a plant produces seeds (has pollinators in the area?), the mean seed number rises continually as a function of plant size all the way to the largest sizes. Recall from figure 5.9 that these plants above Size Class 45 are pretty rare in natural populations, so we don't often see plants loaded with berries in the wild.

Totally obscured by the curves in figure 5.10 is the high degree of variability in all of these functions: variability between populations but also within them. These are just averages across all observations in the decade from 2004 to 2014. Some of the variability in behavior of individuals as a function of size class can be seen by examining numerically what happens from one year to the next; these are the transitions shown by arrows in our life-cycle diagram (fig. 5.8).

Unfortunately, since this set of numbers would produce a table with forty-nine columns and forty-nine rows, a magnifying glass would be needed to pick out the actual numbers. Rather than provide that excruciating level of detail, I will simply reveal that the most important takeaway is that there is a lot of variation among plants in performance over a year.

CALCULATIONS

The population matrix is useful not only for showing the variability in individual plant fates; it is also a complete accounting for the changes that we need in order to do projections of population growth rate. This can be seen by thinking about a given row. If a row is all the plants coming *to* a class *from* many different size classes, we should be able to calculate the actual number that will be in that class the next year by just multiplying the number in the class they are coming from by their chance of landing in that row the following year. Then if we sum all those multiplications across all the classes coming into that class, we will have the total number of plants in that class in the second year. Do that for all rows, and you have projected the growth of every class from one year to the next. By summing the totals across each row, we know how the total population size has changed. This is a lot of multiplication to do, and I don't recommend you try it by hand, but computers are great at this sort of thing, so let's let them do it. Finally, if we do this set of calculations enough times, the growth rate settles down to a single number—this is our number that varies around 1.0 (steady state, no growth population). If it's bigger than 1.0 the population is growing, long term; if less than 1.0, it is declining.

So, the answer for ginseng is (drum roll, please) . . . 0.9835. This feels a bit like the answer ("42") to the question on "life, the universe, and everything" in Douglas Adams's *The Hitchhiker's Guide to the Galaxy*. It seems to come out of nowhere, and if it truly captures something fundamental about ginseng, what exactly does it mean? The next several chapters will supply the context for 0.9835, and, from that context, we will learn a lot about the universe as seen from the point of view of a ginseng plant, growing optimistically on the forest floor. For now, we can see that the result of our monitoring in chapter 4 is supported by our modeling in chapter 5: ginseng *is declining at a rate of about 2 percent per year* across our 30 populations. If I had to place a bet on whether this number is too low or too high, I would say it's a bit too high because the losses of seeds from the seed bank may be underestimated since our seeds were protected by seed cages. However, the general concordance of this number with our monitoring numbers suggests we can have good confidence in estimating a long-term 2 percent to perhaps 3 percent annual decline in our natural populations. Things are not looking good for America's most famous medicinal plant in the wild. But why?

CHAPTER 6

The Root of the Harvest Problem

> The nation behaves well if it treats the natural resources as assets which it must turn over to the next generation increased, and not impaired in value.—**THEODORE ROOSEVELT**

For the past twenty years, I have been fortunate to have a charismatic plant like ginseng to attract talented undergraduate and graduate students to my lab. But the first graduate student to wrestle with ginseng problems was the one who initially acquainted me with the plant. Martha Van der Voort worked at the nonprofit Conservation International in Washington, D.C., and helped fund important research around the world. After a few years at this job, she began to feel strongly that she needed a better firsthand understanding of how science was done in order to judge what should be funded. So her good friend Brent Bailey, who was enrolled in a PhD program in the Wildlife Department at West Virginia University (WVU), convinced her to leave the big city and begin a master's program in Morgantown under the guidance of Dave Samuel, a renowned wildlife biologist and bowhunter. Brent and Dave convinced Martha that harvesting deer was part of an overall cycle of harvesting from the woods that formed tight bonds between Appalachian residents and their surrounding forest. They also convinced her that we knew very little about most of those harvested species and that many of them were plants. Intrigued, Martha set out to quantify the wild-harvested plants of a wilderness area in the Monongahela National Forest. They needed a botanist on her advisory committee, and they found me.

I was astonished to learn how many plant species were being harvested from the wild woods of West Virginia—not only ginseng but also goldenseal, black cohosh, bloodroot, and "log moss" by the boxcar load. Meanwhile, Brent was discovering with detailed surveys of West Virginia households just how widespread this connection to the forest was. Besides the usual hunting and fishing, ginseng digging was most commonly mentioned. Morel mushrooms, ramps, and squirrels were in the mix. But everyone at least knew about ginseng.

Tales of large, human-shaped roots worth a fortune were told. The difference between ginseng and ramps or morels or deer or fish was that they were not generally being harvested for personal consumption but for sale to local dealers who were plugged into the medicinal plant trade. Myths were propagated of distant family members who were so good at finding ginseng they would return from a day's rambling over the hills and hollers with several hundred roots in their bag. Knowledge of locations of favorite ginseng patches were guarded as carefully as whereabouts of favorite fishing holes. The cultural significance of ginseng was written on the landscape, with names such as "Ginseng Mountain" and "Sang Run Road" alongside names on maps with the words "Chestnut," "Beech," and "Walnut."

Where the Wild Ginseng Are

In the Monongahela National Forest, Martha came across ginseng now and then in her randomized plots, but it was never abundant. She was surprised how often it showed up, given that it was supposed to be rare. But maybe this pattern was only true of the mature forest found in wilderness. This prompted us to expand the randomized sampling approach to a larger area—the state of West Virginia, with a few forays into surrounding states of Maryland, Ohio, and Pennsylvania. We sampled private property (including private preserves), state forests, state parks, national forests, and wilderness areas. We tried to make no presumptions about where we would find ginseng, with two exceptions: (1) we only sampled forested areas, under the assumption that ginseng required some shade, and (2) we avoided thickets of rhododendron (great laurel) because the entire herbaceous understory of these deeply shaded places tends to be barren. In all, we combed every square foot of seventy-two acres at sixteen sites spread across the region. Given that every acre contains 43,560 square feet, this took a long time.

Prior to our study, ginseng was expected to be found on north-facing slopes in "cove forests." These are mature forests in isolated bowl-shaped valleys with a diverse tree overstory and herbaceous understory. Instead, we were astonished to find ginseng on all aspects and at all elevations (fig. 6.1). If any pattern emerged, it was that higher densities were to be found on east- and west-facing slopes in this region, though the pattern depended on the elevation. But after the better part of a summer sampling with a field crew of several people, it became clear that there was also tremendous plot-to-plot variation, even within

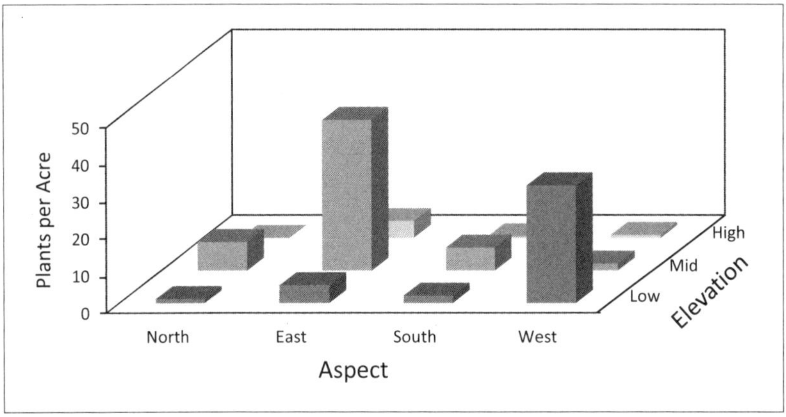

FIGURE 6.1.
Density of ginseng plants depending on slope face (aspect) and elevation

classes of sites. Only one other pattern was obvious—ginseng was more abundant on private lands than on public lands. State parks in particular were nearly devoid of ginseng. Why, especially given that harvesting was not allowed in West Virginia state parks?

In retrospect, perhaps we should not have been surprised by these results. If the historical picture of ginseng harvest painted in chapter 2 is correct, perhaps ginseng has been harvested out of the best sites, and where we find it more often now is in suboptimal sites. Or were west- and east-facing forests actually optimal in the central part of the range? There was really no way to know without doing experiments. This shows the limitation of purely observational science. Even if our sampling was nearly random and ideal in a statistical sense, we could not interpret the findings as unambiguously showing where ginseng grows best because of the long and complicated history of interaction with people.

Since our early study, others have attempted to clarify where ginseng is found and where it is harvested most often. Taking a broad-brush approach using county by county data, Schmidt and colleagues studied factors influencing habitat suitability and harvest across ginseng's range. They found that ginseng preferred higher elevations, cove-shaped land forms, and somewhat basic soils, especially in the south. Relative levels of poverty and road densities both predicted increased harvest.

Harvest and Population Decline

Martha recognized early on that the people part of the equation was the key to understanding harvest impacts, and she set out to quantify the effects of different kinds of harvesters on natural populations. The early pioneering work on ginseng harvest had suggested that harvest intensity mattered, at least at the northern edge of the range in Quebec. Charron and Gagnon's demographic modeling at University of Quebec in Montreal suggested harvest was sustainable only in "good" years, and even in those years a maximum of 16 percent of plants could be taken. This work was expanded by Patrick Nantel, with Andrée Nault and Daniel Gagnon, using the same data to predict the chance that populations would persist for one hundred years under various harvest regimes. They showed that virtually all populations in Quebec were too small to support any harvest at all. But not all harvesters behaved the same way, and Quebec was far from the middle of ginseng's range. Maybe ginseng would be much more resilient in the face of harvest in a more optimal environment.

Martha used our first adopted wild ginseng populations to form models similar to those of Danielle and Daniel but now in West Virginia, near the range center. She based her models on six populations with three years of data for each. At that time, the ginseng harvest season legally began on August 15 in West Virginia (it has since been changed to September 1, in part due to her work). To examine harvest effects, she invented three harvester types, labeling them (1) stewards, (2) compliant (those complying with existing laws), and (3) noncompliant ("outlaws"). Stewards fit the model of Art Logan and his granddaughter Ellie (chapter 2). They go above and beyond the legal requirements to ensure the long-term welfare of ginseng, with specific strategies to ensure replacement of harvested individuals. Stewards delayed harvest by two weeks to ensure berry ripening. They also self-limited harvest intensity of the population, taking only 25 percent of the small and large adults and only those with berries containing seeds. On top of this, they carefully planted the seeds at the optimal depth of three-fourths of an inch, deep enough to hide them from seed predators and prevent them from washing away but not so deep that the seedling would struggle to reach the surface after germination. Compliant harvesters followed the letter of ginseng laws; they harvested on the first day of harvest season, they harvested only plants with three leaves or more, and they left the berries on site by scattering them on the surface. The noncompliant harvester ignored harvest season, digging plants on June 15 before seed production occurred for that year. Also, they removed 25 percent of undersized two-leaf juvenile plants, as well as three-leaf and four-leaf adults. At the time, we knew these

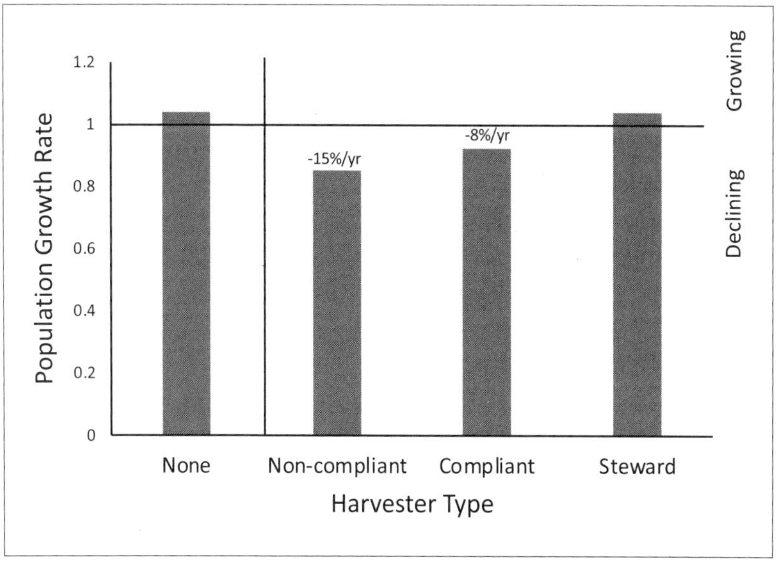

FIGURE 6.2.
Response of population growth rate to three harvester types, compared with no harvest, based on models of two transition years of data in six populations in West Virginia

three types of behavior occurred from reports of dealers and household surveys, although the proportion of each type of harvester was not known.

We probably did not need a fancy population model to tell us that a noncompliant harvester would have devastating effects on ginseng (see fig. 6.2). With a decline of 15 percent per year, it would not take long for harvest pressure to cause extirpation of a population. However, we were unprepared for the degree to which even compliant harvesters threatened these populations. Martha's analysis went further. She was able to dissect the reason for the differences observed, and, in the case of the compliant harvester, the loss of reproductive adults from the ginseng population was the primary cause of population decline. Harvest resulted in elevated mortality of small and large adults, and seed production would eventually decline as well if that continued, resulting in a population crash.

Only the steward harvester would have no impact on population growth rate. While the increased mortality caused by a steward's harvest was a negative, Martha's dissection of the reason there was little difference from unharvested populations showed that the increased seedling recruitment resulting from careful seed planting compensated for the harvest loss. Key to that compensation were the fact that seeds were more mature by September 1 and the fact that they were planted rather than scattered on the soil surface. But one big unknown remained. What percent of diggers were noncompliant, compliant, or stewards? This would determine the ultimate implications of Martha's findings, but in the early 2000s we had not yet observed enough harvest events to make any generalizations. I will return to this question in chapter 9.

Coincidentally, Martha's analysis suggested two key ways in which ginseng regulations could be improved. While it might be impossible to enforce a requirement that harvesters plant seeds at an optimal depth, that could at least be a recommendation. Secondly, perhaps existing harvest seasons were not optimized. Different states at that time had a wide variety of season onset dates, ranging from August 1 (Pennsylvania) and August 15 (West Virginia, Virginia, Kentucky, Tennessee, and Georgia) to August 20 (Vermont, Maryland) and September 1 or later (other states).

The Science of Harvest Seasons

The haphazard array of harvest season dates led us to ask whether there was any rhyme or reason to the differences. Ostensibly seasons were set in order to ensure seed maturation prior to harvest, but were fruits really ripening by August 1 in Pennsylvania while one state away in New York, they waited until September 1? We had our doubts, but what was really happening on the ground? To find out, we cobbled together a set of fruiting monitors from across ginseng's range, from North Carolina to Missouri up to Quebec and Maine. We agreed to monitor berry ripening stage on August 15, September 1, and September 15.

What we found backed up our hunch that harvest seasons were not based on good science (fig. 6.3). On August 15, only the two Missouri populations had 100 percent ripe berries, while berries in most populations were entirely green (unripe), with undeveloped seeds at that date. By September 1, the majority of populations showed more than 50 percent ripe red berries, and even the berries that were green were mostly large and well-developed. By September 15, only three populations had more than half green unripe berries in the population. Clearly, any harvest season beginning prior to September 1 was reducing the potential for harvesters to ensure replacement of the plants they dug.

Beyond this broad pattern, we found no basis for differences among states in their harvest seasons. On the face of it, shifting to a September 15 harvest season onset date might have seemed the obvious regulatory solution. However, we identified three negatives to that large a shift in the season onset. First, by that date, many berries would have already dispersed, making it difficult or impossible for a harvester to plant the seeds. Remember, one of the big benefits

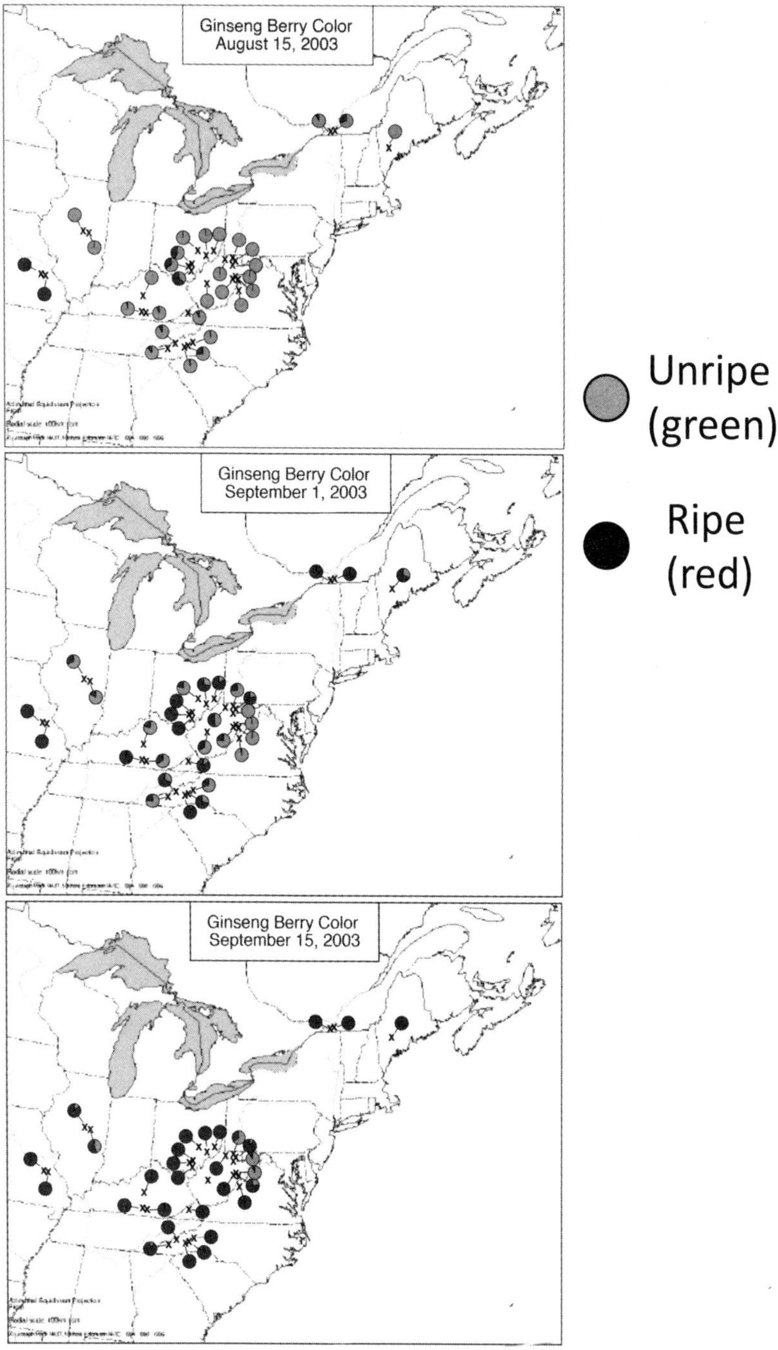

FIGURE 6.3.
Geographic pattern of berry ripening. Each pie chart shows the proportion of berries ripe (red) or unripe (green) at each date.

of the steward harvester in Martha's models was the fact that they planted seeds at an optimal depth, which ensured better survival and germination from the soil seed pool and thus replacement of the harvested plants. Second, a September 15 deadline would require legislative action in eighteen of nineteen states with ginseng programs, which could slow implementation. Finally, harvesters might worry that ginseng plants would be senesced or browsed by that time and thus be more tempted to harvest out of season, which would encourage more noncompliant behavior. In the end, September 1 appeared to be the best compromise. A uniform harvest season among states would also have the advantage that harvesters would be less inclined to cross state borders to sell roots in a state where the harvest season started earlier.

The data from this paper, including complementary studies on germination at different phases of seed ripeness, convinced the U.S. Fish and Wildlife Service to pressure states to modify their ginseng programs accordingly. In a happy instance of science driving policy, gradually states pushed through these changes, and now we have a nearly uniform harvest onset date (September 1) across ginseng's range. This seemed like real progress. On top of this, states were including recommendations for planting ripe berries near the place where the harvester dug the parent plant. We had reason to hope that most harvesters would now be behaving more like stewards of our common resource than exploiters who were depleting the resource.

In our early years of twice-yearly censusing, we noted plants disappearing between our spring and fall censuses. Our first thought was that there was a lot of out-of-season harvesting going on. But in most cases there was little evidence of digging at the spot where the plant was located. Plus, when we returned the following spring, these plants often reappeared, so that ruled out harvest. Clearly something was happening to the aboveground part of the plant between spring and fall. We needed a ginseng sleuth to get on the case.

CHAPTER 7

The Case of the Disappearing Plants

> Since then I have lived to see state after state extirpate its wolves. I have watched the face of many a newly wolfless mountain, and seen the south facing slopes wrinkle with a maze of new deer trails. I have seen every edible bush and seedling browsed, first to anemic desuetude, and then to death.—**ALDO LEOPOLD**

"How often do you find that ginseng skips years?" This is the question that most often follows when I reveal my downward-slanting N versus time graph (chapter 4) in public forums. Why? I'm sure many in the audience think the decline is an artifact of these "out of sight" plants. And, to be fair, to those who know plants, this notion is not as strange as it sounds. Orchids, for example, are known for remaining dormant for one or more seasons after appearing aboveground.

Thousands of diggers set out to find ginseng each year, and they generally do so in the fall. Or at least they should, since harvest season doesn't start until September. Also, rarely do diggers hunt randomly. Like an angler returning to a favorite river pool, they are often returning to spots where they have seen their quarry in prior years. The spatial memory of some diggers can be astounding as they clearly recall a particular rock or log with a certain plant next to it, even though they may not have been back to that spot for a year or more. Sometimes they retrace their steps specifically to dig that plant. Of course, today that memory may be assisted by GPS on a smartphone. By whatever means a repeat visitor tracks down a given plant, it is not uncommon for the plant to be gone. They tuck this away in their memory, noting to check again next year because in the past some ghost plants have reappeared. Indeed, references in the literature to ginseng "hiding itself" are frequent. It is always hard for people to question what they have seen with their own eyes—a plant was there one year, missing the next, but back again the following year.

"Very rarely, if ever," is my response to their question ("How often does ginseng skip a year?"), and as you can guess, this declaration is met with intense skepticism. The explanation for the misconception that ginseng commonly skips seasons is fairly simple. If they had been there at the season's beginning, they would have observed their plants, emerging from winter dormancy as usual. The root was there all along, of course, but so was the top, at least at the beginning of the year. Something did indeed cause the top to disappear during the summer. We too had observed plants going missing, but our spring censuses told us that in the vast majority of cases these plants were not skipping seasons.

Whodunnit?

Enter Mary Ann Furedi, graduate student and plant detective. Mary Ann joined the my lab after jettisoning her career as an forensic scientist involved in investigating headline-producing murders. A highlight of her budding forensic career came when a photo of her appeared in *People* magazine, loading a corpse into a station wagon to be taken for an autopsy. Despite the glamour of her professional job after her undergraduate degree, Mary Ann longed to be outdoors, and she applied to graduate ecology programs where she might combine her love of nature with a career. She enrolled as a master's student in the Department of Biology at West Virginia University. Intrigued by her innate talent for sleuthing, I suggested that she investigate the case of the disappearing ginseng plants.

As her excitement grew in her quest to resolve the case, it became clear that Mary Ann's project would require attacking the problem from several different angles and that it was likely to take four to five years to answer just some of the questions that quickly began gnawing at her. In other words, it was a PhD project, not a master's. But I was confident that Mary Ann was PhD material, so I was all for it. Mary Ann switched degree programs and launched several lines of inquiry right away.

The first question: What was causing ginseng plants to disappear between our spring and fall censuses? Mary Ann thought the chance that she would catch the culprit in the act with a firsthand observation was slim. So she took advantage of a relatively new technology for the year 2000—motion-sensitive game cameras. Though many sophisticated models can now be had for less than $100, at the time such highly specialized cameras were mainly used by researchers and were very expensive. Mary Ann wrote a grant to purchase such a device and quickly set one up to be a passive observer next to a small patch of ginseng plants.

The second approach Mary Ann used to pursue the culprit was to select seven populations to visit every three weeks for an entire summer. This entailed

FIGURE 7.1.
A young buck approaches a four-prong ginseng plant, and within three minutes all of the leaves are gone, leaving characteristic torn petioles behind. Note the time stamp on the images showing three minutes elapsed between frames.

following the entire "phototrail" (see chapter 4) through each population all summer and into the fall. While there, she had a detailed list of clues she checked at every plant to learn its fate. This part of her research required her to don hiking boots, desert tan hiking pants, tie-dyed T-shirt, and ball cap, and then drive all over the state to visit the isolated and often-steep forested sites. In the process of these solo excursions, she encountered rain and hailstorms, poison ivy and thorny greenbrier, charging black bears, and rattlesnakes the size of your arm. She loved it.

Game cameras were the ticket. With this technology, in short order our suspect was caught in the act of removing the top. Long before this confirmation, we kept encountering stalks that had been nipped by something. We strongly suspected white-tailed deer from other circumstantial evidence as well—telltale tracks, deer poop, and a pattern of damage that has been described for other species. But we didn't really know for sure. Since Mary Ann's first photographic confirmation, however, we have since obtained many more images of deer munching on ginseng plants (fig. 7.1).

We don't believe ginseng is a preferred food plant for deer. The little buck in figure 7.1 walked on and left the large plant in the foreground untouched. In fact, if you look closely, you can see a small two-leaf ginseng plant underneath the browsed plant. The deer may have removed one of those two leaves but left the other one. Mary Ann observed this a lot in the field. Quite often just a few plants out of a larger group would be browsed, and some plants were only partially browsed. Perhaps this is unsurprising since ginseng leaves have a slightly bitter taste, at least to some human palates.

Browse Rates

Just how big a problem are deer, and how does it compare with human harvest? To judge, we need to take a look at deer browse rates in our thirty populations. If we had five thousand excellent game cameras, one for each plant in our census, we would not have to estimate. But that is unrealistic and way too expensive (and maybe the cameras would alter the browse rate), so instead we need to estimate browse rates. The first approach we can take in such estimation is to count the number of plants that show missing leaves like those browsed stalks produced by our buck in figure 7.1. The problem with this estimate is that Mary Ann's frequent visits to six populations showed that browsed stalks generally fall over and start decomposing after about three weeks. Therefore, we know that for a population we check only twice per year, we are only sampling about six weeks of a fifteen-week growing season. Knowing this, we can count the browsed stalks at our two census times and multiply by 2.5 (15/6) to get our first estimate of browse rate. This assumes that the browse rates three weeks before our two censuses each year are not different than browse rates for the rest of the year. It also assumes that the three-week estimate for stalk presence after browsing is correct (this could vary a bit) and that a fifteen-week growing season is standard (some may be longer than this).

A second estimate can be obtained by counting missing and browsed stalks by the time of the second (fall) census and making the assumption that all missing browsed stalks are caused by deer. This could underestimate browsing because there is more time for stalks to be browsed after the fall census (and we wouldn't know, from these numbers), but it could overestimate browsing because it attributes all missing stalks to deer effects (and there may occasionally be other causes). Nevertheless, because we know deer are probably the biggest reason that stalks go missing, the estimate is probably not too far off. We use our two browse estimates together for the thirty populations to see what the range of browse might be.

The first thing we noticed was that the two browse estimates showed a strong positive correlation, suggesting they are estimates of approximately the same thing and therefore capturing important differences among populations. The average browse rate lies between 21 and 28 percent of all plants. In other words, about a quarter of all plants are browsed each year. No wonder diggers think ginseng skips years. But the range of values is also interesting. The lowest browse rate we found was between 1 and 12 percent, while the highest was 44 percent by one measure and 67 percent by the other. No population was completely immune from browsing. Susan Farrington, who performed an excellent major study of deer effects in Missouri, showed similar, though slightly lower, average rates of browse (11–33 percent; mean of 17 percent) in her populations.

Another important feature is that browse is not the same across all size classes. Mary Ann showed that deer targeted large plants, including those that were most likely to produce seeds. This meant that while only about a quarter of all plants were browsed, these were the most important plants in terms of propagating the population. We think deer target these large plants for the simple reason that deer are just being efficient as they forage through the woods. If they can get a larger reward (more leaves) without stooping as low to the ground (one-leaf plants are shorter and harder to reach), then they save energy.

Mary Ann also showed that most browsed plants survived, but they often shrank in size (both height and leaf area) the following year. In another subtle twist, the effect of browse on plant shrinkage was greater when browse occurred earlier in the year. This makes sense when you realize that plant leaves are its feeding organs, absorbing light and converting the energy of light along with carbon dioxide to form sugars and, later, more complex molecules such as starch. Starch is the plant's way of storing energy and mass to support growth of new roots, stalks, and leaves the following spring. By removing leaves from the plant early in the year, deer essentially starve the plant that year, preventing it from preparing for next year's growth. If you think of the storage root as a pantry of goodies to support the living plant's own growth, then, by removing leaves, *early*-browsing deer prevent the plant from restocking the pantry shelves at the worst possible time, when the plant has just emptied the pantry by producing a new top. In contrast, if a deer browses that plant in September or October, the pantry is full, and it will have little effect on future size or reproductive output. Some plant species could better deal with browsing like this by regrowing a new top in the same year that browsing occurred. But ginseng, and indeed most understory plants, do not have sufficient energy to do this, and they instead form a new terminal bud belowground, waiting until the following year to put up a new top. Farrington's Missouri plants showed the same pattern of size-dependent browse and time-dependent effects that Mary Ann Furedi showed in her work. With repeated browsing (two or three years in a row), ginseng plants can lose so much energy that they die.

Are Deer Always Bad?

Can deer have any positive effects on ginseng? I can think of three possible ones. First, if deer not only eat the leaves but also eat the ripe fruits of a gin-

FIGURE 7.2.
Mary Ann Furedi collecting deer pellets after a feeding trial with caged white-tailed deer. A total of eight deer were fed, including four yearlings, two does, and two bucks. Several thousand seeds were fed to the deer over a period of three weeks.

seng adult, they could serve the function of dispersing seeds. We had never seen deer poop with ginseng seeds in them, but that is not proof one way or another. Again, Mary Ann was going to find out using a more scientific approach. She located a game farm with captive deer, then fed ripe red ginseng fruits from a wild-simulated ginseng farm to the deer. Finally, she gathered the feces in the ensuing days for analysis (see fig. 7.2). Pellet samples were carefully sieved to look for intact viable seeds.

The bottom line was that deer adored ginseng fruits, often selectively picking them out of the corn mixture of their standard feed. However, from the thousands of fruits fed to deer, no seeds survived passage through the gut. Deer were predators. This was a little surprising because other researchers had discovered that seeds of dozens of plant species were dispersed by white-tailed deer. But most of those had relatively small, hard seeds. Ginseng seeds are apparently just large enough, or soft enough, to be destroyed by the rough journey through the extensive digestive tract.

A second possible benefit of deer is greater browsing of ginseng's competitors than of ginseng itself. On this, we do not yet have a definitive experiment to tell us whether the competitive release experienced by such preferential browse of other species compensates for the detrimental effects of browse itself. However, circumstantial evidence suggests this is not the case. We'll revisit this issue below in talking about the special case of ginseng growing on tops of rocks unreachable by deer, but suffice it to say, despite the fact that ginseng is not particularly a preferred species, many populations appear to be in decline because of overbrowsing, despite a lower amount of competition from neighbors. Finally, ginseng does not seem particularly responsive to neighboring plants, a fact we discovered when we examined invasive species effects. If there is an effect of competitor reduction, it is a temporary benefit that eventually goes away as ginseng becomes one of very few understory plants remaining for deer to eat after chronic overbrowsing.

A third possible benefit of deer is the effect that removing the top is expected to have on apparency to human harvesters. Clearly if harvesters follow the rules and wait until September 1 to begin digging roots, if a quarter of adult plants are missing but still alive belowground at that point in time, deer browsing will have prevented deaths of those missing plants. If the stalks are gone, harvesters will have no way of detecting that the roots are there. Alas, perhaps this is the only way in which overabundant deer have beneficial effects. Farrington's research in Missouri suggested as much. Unfortunately, it may be an example of one negative effect reducing the severity of another negative effect; wild ginseng would still be better off without either.

Effects on Population Growth

Mary Ann Furedi's pioneering work also examined the demographic effects of deer, using a clever approach of modifying the predictive models like Martha Van der Voort had done before for harvest. Using seven populations from north-central West Virginia, she constructed models based on all the plants in the populations, or, alternatively, based only on the unbrowsed portion of the population. This allowed her to simulate the effects of removing herbivory on population growth rate without actually having to erect eight-foot-tall fences in the field.

Across the seven populations, Mary Ann found that when browsed plants were included in her data set, populations were declining by 3.7 percent per year (on average). However, if browsed plants were removed from the data set, the populations increased by an average of 1.1 percent per year. In other words, deer were making the difference between an increasing, healthy population and a declining, unhealthy one. Comparing this to Martha Van der Voort's harvest study, we see a major difference—Martha's "compliant" harvester reduced population growth rate much more than deer (−8 percent per year vs. −3.7 percent per year). However, not all populations experience harvest, and even those that do rarely get harvested every year. Deer, on the other hand, are omnipresent, and rates of deer browse far exceed rates of harvest.

Farrington's Missouri study again showed a comparable magnitude of deer browsing effect, with a simulated unbrowsed population enhanced by 2.9 percent relative to the ambient browsed population. The slightly smaller difference than that found by Furedi is easily explained by the slightly lower ambient browse percentages in Missouri.

Effects on Population Viability

In the same way that we can simulate the effects of deer browse on population growth, when we have multiyear data sets, we can perform population viability analyses (PVAs) that reveal the long-term conservation impacts. Mary Ann's demographic data proved useful once again for this exercise. We found that if there was no reduction in deer browse rate from current levels, then a viable population that would persist for a century would have to contain at least eight hundred individuals. None of the thirty populations we studied came close to that number. The greater the reduction in simulated browse rate, the greater the population viability became at any given initial population size. Even so, with a 75 percent simulated reduction in browse, about half of our thirty monitored populations would still not achieve viability in terms of their initial population sizes.

Our population viability study made a significant news splash because it was published in the nation's top scientific journal, *Science*, and because people had always assumed that human harvest was ginseng's biggest threat. This study suggested maybe that wasn't the case. The harvester of greater import had four legs, not two. This didn't absolve the two-legged mammal, because people were responsible for regulating deer populations. But it did add an important layer of complexity to our understanding of what was happening to ginseng in the wild.

What if There Were No Deer?

Taking the big picture view, a state of no herbivory in a habitat is undesirable, in that animals that eat plants—herbivores—are the second layer of a healthy, multi-trophic ecosystem. However, as an extreme measure, one could imagine fencing deer completely out of a ginseng population and observing the results.

FIGURE 7.3.
Photos taken in June 2017 off and on the perched rock in Population 27. The isolated, flat-topped boulder could not be reached by deer, resulting in abundant and diverse forest floor vegetation, including ginseng.

Serendipitously, just such an opportunity offered itself to us in the course of our research, and the fence had been in place for thousands of years. The fence happened to be a huge boulder, ten feet tall, with a flat top (about thirty by fifty feet) that supported a lush plant community, including trees. In that plant community was a small ginseng population I will call the "on-rock" subpopulation. Off the rock, deer could reach the plants. I'll call that the "off-rock" subpopulation. The beauty of this particular large population is that we had been studying it since 1998, so for twenty growing seasons we could observe the relative dynamics of the off-rock and on-rock plants. In the photos in figure 7.3, you can see how huge the differences were between the habitats just a few feet away from each other. What did we find about ginseng growing in these two places?

The greatest difference in average characteristics of plants on and off the rock were their observed growth rates. We measured this in terms of relative change in leaf area from one year to the next. A plant such as ginseng, when healthy, would be expected to grow more leaf area each year, so the relative growth rate in leaf area should be positive, at least on average. However, the two subpopulations were very different in this regard. The on-rock subpopulation was indeed growing, but the off-rock group was slightly shrinking, on average. But we found very little difference when comparing on-rock and off-rock size-dependent survival or reproduction. This can be explained by the huge effect that deer have on the ability of ginseng to store reserves to support growth the next year. Note that the average for the off-rock plants includes the approximately three-quarters of that population that is *not* browsed in a given year; so while those unbrowsed individuals may have grown, this was more than compensated for by the devastating effect on the browsed plants.

What does this change in individual growth rate mean for overall population growth rate? Recall we determine this useful statistic with our population models, so that is exactly what we did, separately, for the on-rock and off-rock subpopulations. The models predicted what you might expect. The off-rock subpopulation was declining at a rate of 4.5 percent per year, while the small subpopulation on the rock was increasing at a rate of 3.7 percent per year. What is more, when we plot what has actually happened on the rock, these model predictions are borne out very closely. Perhaps even more startling than the population trend was the change over time in the importance of the little group of plants on the rock top in Population 27. In 1998, when we started this study, the on-rock group of plants accounted for 3 percent of the plants and only 2 percent of the seeds in the entire population. By 2017, these plants accounted for 9 percent of the entire population and 37 percent of the seeds!

Was this population an outlier with respect to deer browse? Did I purposely select the worst-case scenario to illustrate potential effects of deer? Unfortunately, the answer is no. I do wish I could say this population was abnormally high on the browse scale, but in fact it is smack dab in the middle of the range shown by the group of thirty. This should really give us pause because if we flip back to the relative population size versus time chart in chapter 4 (fig. 4.5), we see lots of populations trending downward just like Population 27. Is this due to deer? For that answer, we would need to do a more synthetic analysis of the big data set. For now, we can conclude it is most certainly one major contributor.

We learned one more thing from our twenty-year study of Population 27: there is a clear signal of overbrowsing by deer even with just one year of data. That signal has to do with the size structure of the population. The subpopulation off the rock is overloaded with two-leaf plants that we typically refer to as juveniles, whereas the protected group on the rock has lots of large reproductive adults and new seedlings. The explanation is clear from the growth data; browsed plants regress in size, often from three-leaf or even four-leaf plants down to two-leaf plants. They may still be quite old, but they are stunted. In a healthy population, two-leaf plants represent a transitional stage between one-leaf seedlings and three-leaf adults as they grow. Typically, a plant might stay in that stage between one and five years, depending on the quality of the environment. But in a heavily deer-browsed population, two-leaf plants are a mixture of true juveniles and stunted adults. When we look at all thirty populations in the big data set, we see that the proportion of two-leaf plants is a pretty good predictor of browse rates.

Although Population 27 is in many ways a typical population of ginseng with respect to deer browse, it is worth looking at an extreme case just to see how bad it can get. For this, we look at a Virginia population (Population 17) that is well protected from human harvest. Unfortunately, it is not well protected from deer. Walking through this population year after year, one just has to marvel that it is still there at all since virtually all of the plants there, even the rare three-leaved plants, are stunted. More than half the plants (57 percent) in that population are two-leaf stunted adults. Even more amazing is that over fourteen growing seasons, a total of only forty-two berries were produced (three per year) with a total of sixty-five seeds (less than five per year). This was for a population of over two hundred plants! The fact that ginseng has hung on in the face of that onslaught is a minor miracle, and it attests once again to the resilience of this species in the face of heavy stress. The 57 percent decline observed over just one decade, however, suggests that this population may not be around much longer.

Assigning Blame

I almost stepped on him. With eyes trained on the ground, looking for ginseng, I hopped over a downed log and jumped aside when I realized what was in front of me. Scrunched up against the log on the far side was a tiny newborn fawn, with stripes of dots that looked like sun flecks. She remained frozen, though I was only two feet away. Her huge eyes and wet nose captured me in that moment the way a human baby does, all innocence and possibility.

"Did you kill it?" an esteemed scientist colleague asked me, when I described this scene from our spring censusing trip. The question brought me up short. How could I possibly even contemplate killing such an adorable wild animal? But I knew where he was coming from. In fact, the question was perfectly logical from a plant conservation standpoint because that small fawn was about to embark on a lifetime of plant consumption.

"No, I couldn't" was all I could muster. Only later did I realize that this had been a weak response. It wasn't just that the fawn in question was undeniably cute. The truth is, I could not blame any individual deer for doing what any live animal must do: eat to survive. The fact that there are too many deer is the problem. That problem has distinctly human causes.

FIGURE 7.4.
Fawn blending in with the forest floor

To differentiate scientific evidence in the book from my opinion, I have included boxes labeled IMHO (In My Humble Opinion) in which I express my personal opinion about an issue, setting it aside as just that, not scientific fact. The first of these appears below. This device evolved from my Aldo Leopold Leadership Program training, in which we were advised to separate the science from the clearly opinion portions of our outreach writing. In other words, scientists are entitled to have opinions but need to clearly articulate when they are giving them.

IMHO

Deer are mismanaged in this country. Or, rather, they are managed well, but primarily for the benefit of hunters, with little regard to other users of the ecosystem. Although I am oversimplifying here, this is a relatively straightforward consequence of the management paradigm that has been set up. In most states, fees from hunting licenses are used to manage deer. On the surface, this may seem logical. However, it sets up perverse incentives that are bound to have unintended consequences. In order to have adequate resources for deer management, wildlife management agencies encourage hunting by maintaining high populations of deer to hunt. However, as we have shown with detailed data on just one understory plant, the consequences of high deer populations are devastating over the long haul. This in turn affects the resource availability for all other species that are supported by that understory growth. In addition, deer are not confined to the woods, so they move into adjacent lawns and gardens, with costly effects on people trying to grow plants. They also cross roadways, often in an unpredictable fashion, causing a large toll on cars (over $1 billion per year), human injuries, and about two hundred human deaths per year. We all pay for this with elevated car insurance rates. Compare this to the human deaths caused by rattlesnakes (about five per year), grizzly bears (two), mountain lions (less than one, on average), and wolves (zero). Clearly, deer management will not change due to deer effects on ginseng alone. But as part of a larger picture of the negative effects of overpopulation, deer browse on ginseng contributes to the case that we need better strategies for lowering population sizes. This could include simple things like expanding hunting seasons, promoting doe hunting, and reintroducing top predators, especially where hunting is inadequate to the task. But as long as deer are managed in isolation from the other components of the ecosystem, the status seems unlikely to change. The solution? I believe we should be managing whole ecosystems. Metrics for the health of the ecosystem, broadly measured as biodiversity and productivity, should be maximized, with exceptions made for rare or endangered species requiring specific habitats, which may require species-based approaches.

With the case of the disappearing plants solved, we now had two explanations for the decline of ginseng: human harvest and deer browse. Ginseng had two strikes against it. Would there be a third? A clue lay deep in our demographic data, indeed in the data for Population 27, accumulating since 1998. By the mid-2000s, patterns in the data began whispering, and it took a mathematically minded undergraduate to listen carefully enough to hear it.

CHAPTER 8

The Elephant in the Room

> For way too long the politicians and people in power have got away with not doing anything at all to fight the climate crisis and ecological crisis. But we will make sure they will not get away with it any longer. We will never stop fighting, we will never stop fighting for this planet, for ourselves, our futures and for the future of our children and grandchildren.
>
> —GRETA THUNBERG

Climate versus Weather

As I write this, the temperature in Morgantown, West Virginia, is ninety degrees Fahrenheit. Not terribly unusual for this place, but the kicker is the date: October 3. This alone does not mean the climate is changing, even if it is a record. It means the weather today is unseasonably hot for this time of year at this place. It is a data point in an overall pattern, and certainly one that makes you sit up and take notice. Climate, by contrast, is a collection of data points revealing a pattern. It is the average weather at a place, integrated over time. So, whether you are experiencing an unprecedented hurricane in the Bahamas, New Orleans, Houston, or Cape Hatteras, you cannot definitively say that one particular event is the result of climate change. On the other hand, if you observe a statistically significant trend in the number or intensity of hurricanes over time, that *is* a manifestation of climate change. If you observe an increase in average or maximum October temperatures over the past few decades, that *is* climate change. If you observe an increase in the variability in precipitation (from droughts to deluges) at a place, that *is* climate change.

I have spent a paragraph drawing the distinction between climate and weather because when scientists discuss climate it really can be confusing. For example, a reporter may ask, "Was that hurricane caused by climate change?" Knowing the distinction between climate and weather, you can understand why a climate scientist answering that question does not give a simple yes or no answer. After hesitating a moment, they give what seems to be an unsatisfying answer: "Well, the chance of that hurricane occurring is X percent greater due to climate change." Or they may say, "The dice were loaded in favor of this hurricane." The questioner wanted a yes or no and instead got a probability. But the questioner was confusing weather and climate.

The reporter is likely to get an even more unsatisfactory answer if they ask the follow-up question: "Will we get more hurricanes like this in the future?" Answer: "Model simulations show that if we keep increasing fossil fuel consumption at the rate we are now, then the chance of Category IV hurricanes . . . ," and the listeners' eyes glaze over trying to comprehend it. If the scientist had simply said yes, they most likely would have been correct. But scientists are uncomfortable making such generalizations without providing the contextual ifs, ands, and buts. In truth, the answer depends on many unpredictable factors. Perhaps the largest: Will humanity decide it is in our collective best interest to address climate change? Or will we go on with business as usual?

According to many objective measures, using millions of data points, the climate is changing. Global average temperatures are higher. Glaciers are melting. Sea level is rising. Oceans are getting more acidic. These are facts. Scientists also know, with a very high degree of confidence, that activities of humans are the cause. This is a deep topic and one not always easy to understand without training in atmospheric physics, but if you wish to delve deeper, I recommend one of many websites, vetted by scientists, devoted to explaining climate science. NASA has a superb site explaining how we know about climate change: https://climate.nasa.gov/. Follow links at that website to read about the evidence, causes, effects, scientific consensus, and vital signs. If the website has been taken down by the time you read these words, the reasons were political. The truth about climate change can be very inconvenient for some. Better to hide the evidence.

However, this book is not a climate change book; it is a book about how a modest understory plant that a lot of people care about is affected directly and indirectly by human activities. Though we might wish to, we cannot ignore climate change as one of those human-caused changes in the environment and still be intellectually honest. For our studies, as for the broader field of ecology as a whole, climate change is the elephant in the room that we cannot ignore, political inconvenience notwithstanding.

Response to Hot Years

As with climate science itself, one data point, such as one ginseng population's response to one year of drought, would tell us very little. However, as we accumulated census data for thirty populations over more and more years, we began to think it might be possible to observe climate response patterns. To take on this project, I needed someone with good training in mathematics and statistics as well as an understanding of biology. Normally this would mean a graduate student, but one day an undergraduate student from my class, Kristen Hunter-Cevera, came to my office, inquiring about the possibility of doing a capstone project in the mathematics department. She seemed exceptionally calm and mature for her age, so I broached the subject of studying ginseng's population response to weather differences among years. She eagerly accepted the challenge. I first showed her the census data for the population we had been studying since 1998, figuring our longest data set was a good place to start. Then I showed her how to form a transition matrix model to determine the population growth rate (as in chapter 5). Finally, we devised a way to extract temperature data from a nearby long-term database. Kristen then disappeared with her computer, stopping back in for questions rarely. This was a daunting project for an undergraduate, and I began to worry I had overdone it, perhaps discouraging a talented student from ever pursuing research in the future.

I needn't have worried. Near the semester's end, Kristen showed up in my office looking calm and unruffled. She proceeded to show me the results of her analysis. My eyes bulged. She had clearly shown that years with high mean summer daily maximum temperatures were associated with reduced population growth. Only one problem prevented us from publishing that work; we still had analyzed data from only one population, albeit based on nine years of information. She did present the work at a national meeting of the Ecological Society of America, attracting great interest from the poster session attendees. Kristen's project was over, but she was not done as a population biologist. She went on to get a PhD at MIT/Woods Hole Oceanographic Institute, where she formulated matrix models of marine phytoplankton.

"Diligence is the mother of good luck." The quote I wrote in my high school yearbook was a particularly nerdy one from the quotable Benjamin Franklin. Only later did I discover there certainly is dumb luck as well. While Kristen was working on her project, a new graduate student came knocking on my door, wondering aloud about joining my lab. I listened to her pitch. Sara Souther had just gotten out of the Peace Corps with her husband in Paraguay. She was familiar with ginseng because she had been a summer conservation intern in my lab a few years before. Back then, she had seemed to enjoy being outdoors as a summer job and was plenty bright and inquisitive, but I had no inkling she wanted to be an ecologist. But now, with the Peace Corps in the rearview mirror, she exuded a degree of seriousness and resolve I found convincing.

As a general practice, I would recommend my best undergraduate students go elsewhere for graduate training—to get exposed to new schools of thought and hence new ways of thinking about ecological problems. But because I was so excited by Kristen's work, and I knew Sara Souther had the potential to be a phenomenal student, I ignored my rule of thumb and selfishly pulled out the stops to try to recruit her. Fortunately, I had just been funded by the National Science Foundation with a five-year grant to study ginseng and had included a conservation graduate fellowship in the grant. This would allow Sara to do less teaching and focus more on her research while she was getting her PhD. At the same time, I directed her attention to Kristen's intriguing pilot study, hoping Sara could expand the work to many more of our study populations. In the research that followed, Sara discovered a possible third explanation for the decline of ginseng.

When It Rains, It Pours

Sara's findings regarding ginseng's response to precipitation and temperature mirrored the overall patterns predicted from basic ecological and evolutionary theory. A fascinating element was that the graph of precipitation versus population growth confirmed the ecological prediction, while temperature versus population growth illustrated the evolutionary prediction, and the two patterns were not the same.

Ecologists conceive of a species as having a niche—a multidimensional concept basically describing the set of environments within which a species is found, including its role in the community. We're diving into theory for a moment, but bear with me. Part of this theory says that there is a geographical area with conditions that match the physiological capabilities of a particular species—this is the "fundamental niche." Beyond the edges of that geographical area, conditions are too harsh in one respect or another for the species to tolerate. The actual geographical area occupied by the species may be further restricted by interactions such as competition with other species, so the "realized niche" may be smaller than the fundamental niche. Layered on top of this idea is the notion that conditions at the range center are near optimal for that species, while at the margins, conditions are suboptimal. In other words, the geographic range of a species is related to environmental factors as they change from one place to another. Thus, along a gradient of environmental conditions, for example, from low to high precipitation, there will be an optimum somewhere in the middle.

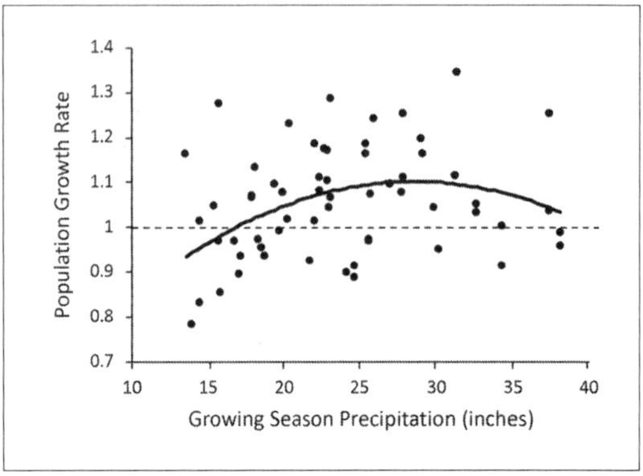

FIGURE 8.1.
Response of population growth rate to between-year rainfall variation in twelve natural populations of ginseng

That is exactly the pattern Sara found for precipitation versus population growth in ginseng. It was not based on geographical variation but instead on precipitation variation among years (see fig. 8.1). The optimum precipitation for ginseng was right about twenty-eight inches of rain during the growing season (April 15–September 30). Also, below about sixteen inches, the survival, growth, and seed production rates were low enough that population size was predicted to decline (population growth rate less than one). Population growth never got below one at the highest precipitation levels, but they were trending downward, suggesting that too much water was not a good thing either.

Rainfall is only one factor among many that determine where a plant will grow, so it is not so simple to look at ginseng's range and point to the places where precipitation is the limiting factor. Nevertheless, noting the pattern of a general reduction in annual rainfall as you go from the top of the Appalachian Mountains westward toward the plains will result in drought being a stressor at the western end more often. At the same time, while proceeding westward, trees begin to drop out, and prairie replaces forest in the Midwest. Since ginseng requires an overstory of trees for shade, the western range edge may be due more to loss of trees than to the direct effect of lower rainfall.

What does any of this have to do with climate change? The prediction for rainfall in a warmer future over much of the eastern United States is that precipitation events and drought events will become more extreme. In other words, ginseng populations, even at the center of their range, will spend more years at the low or high end of the curve shown in figure 8.1. Because the curve is bowed downward at both ends, this in turn means overall population growth rate will be lower. Not good.

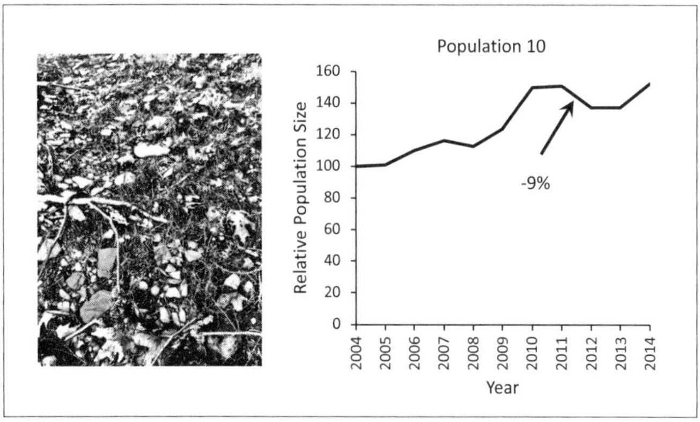

FIGURE 8.2.
Aftermath of Hurricane Irene. The streamside portion of a mature ginseng population in upstate New York was washed away, resulting in loss of 9 percent in an otherwise growing population.

An example of extreme rainfall occurred at one of our long-term ginseng study sites in the Catskills of upstate New York when Tropical Storm Irene dropped about eighteen inches of rain on August 28, 2011. Streams and rivers swelled out of their banks and eroded the soil down to gravel below, leaving tree roots exposed and ginseng roots washed away (see fig. 8.2). That particular population had been growing steadily since monitoring began, but it lost 9 percent of its plants that year to the flood. No doubt many seeds in the soil were washed away as well, but we did not have a good count of those.

At the opposite extreme are drought periods. After several weeks of dry weather, tender herbaceous plants begin to wilt. Ginseng eventually succumbs as well, and ultimately leaves may turn yellow and fall off. The larger plants in the population usually survive this occurrence, but late-season drought is much harder for small seedlings to withstand, particularly if their roots have not grown sufficiently deep. Even large plants have reduced seed production. Flowers and fruits abort in response to drought stress, often resulting in no reproduction that year. Droughts have always happened, but if they happen more often in the future, the negative effects will begin to drag down population sizes.

Thunderstorms and associated hail have long been a part of the summer climate pattern in the eastern deciduous forest, but they can have direct effects as well if they become more frequent. We arrived at Population 2 in Indiana one spring to find the leaves of ginseng as well as other understory plants shredded. We had never seen anything like it. It looked like the vegetation had been

FIGURE 8.3.
Hail damage in an Indiana ginseng population

shot with a BB gun (see fig. 8.3). Then we realized it had been—but the BB's had been hail falling at high speed from a towering cumulonimbus cloud. Local farmers reported crop damage that year as well. Surprisingly, the plants were still present in the fall in the same condition, and, at least in this case, the long-term damage was not evident. However, such tears could easily allow fungal pathogens entry into the leaf, resulting in infection and later death.

In Population 3, a lightning strike from a particularly violent thunderstorm killed adjacent large trees shading a ginseng population. In this case, the canopy opening increased the light reaching the forest floor and ginseng responded with greater growth. One particularly beefy four-prong plant grew over our knees in height. Not all events associated with extreme precipitation are bad.

One more thing was clear from Sara's analysis: ginseng from the hills of western Kentucky, the mountains of West Virginia, and the Catskills of New York all responded to precipitation variation from one year to the next in the same way. There was a broad and relatively uniform "climatic envelope" when it came to rainfall variation among growing seasons.

The Temperature Niche

If ginseng's response to temperature was like its response to precipitation, we would expect a nice bell curve of population growth in response to temperature variation among years. This would fit the ecological model suggesting that northern and southern populations would show contrasting behavior when it came to inter-annual temperature variation. Southern populations would do better in cool years (relative to their long-term norms), while northern populations would do better in warm years (relative to their long-term norms); in other words, they would be on opposite sides of the bell curve, and both would be near their temperature limits.

That's not what Sara found. Instead, each population had its own bell curve (see fig. 8.4). This was puzzling. With respect to temperature, each population was acting like its own species with its own ecological niche! How could this be? Were we dealing with different species of ginseng everywhere we went? What was going on here? This pattern has extraordinary implications in a warming climate, as we shall see.

Understanding this pattern requires some understanding of evolution, but this evolution does not result in new species. At least not yet. But let me back up. In plants, as well as animals that do not move around much, the individuals that are most suited to the local environment wind up leaving the most off-

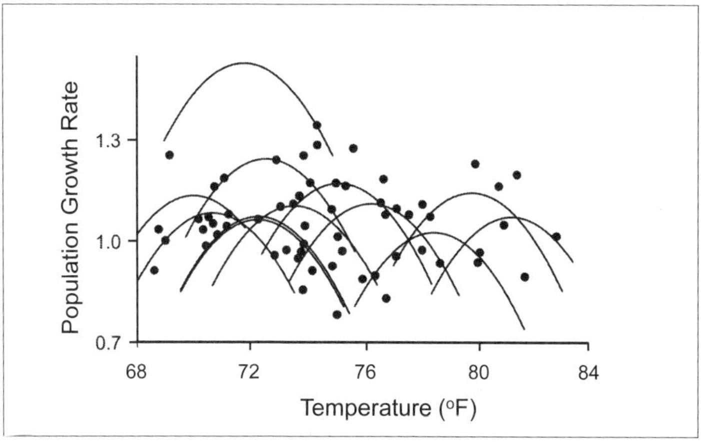

FIGURE 8.4.
Relationship between population growth rate and inter-annual growing season mean maximum temperatures in twelve ginseng populations across many seasons

spring, either because they survive better to adulthood or because once they become adults they acquire the most resources that can be put toward producing offspring. This is nothing more and nothing less than the process of natural selection, first articulated clearly by Charles Darwin in the nineteenth century. In fact, better survival than others in a population only matters in that it results in more offspring, a notion that Darwin understood very well. The saying "survival of the fittest" was not Darwin's and is not the best description of natural selection. If we want to boil natural selection down to a few words, we should really be saying "reproduction of the fittest." And we should define "fittest," otherwise it sounds like we are saying "reproduction of those that reproduce the most," a tautology.

But natural selection is not a tautology. Natural selection is a process whereby physiological and morphological trait variation among individuals affects reproductive success. When combined with the existence of a genetic basis for that variation, the result is evolution over time, and the population becomes better adapted to the environment. The fact that plants are relatively stationary in space means that they can become very specialized to the local environment with ongoing natural selection. The key for our understanding of the many bell curves in figure 8.4 is the notion that the traits that are optimal in northern, cooler climes will be different than the traits that are optimal in southern, warmer climes. Therefore, over time, the genotypes in these places will diverge. They become what biologists call "ecotypes," or locally adapted populations.

As it turns out, for plants, locally adapted ecotypes are the norm. Plants as divergent as the Arctic tussock sedge called cottongrass in Alaska, the dwarf alpine plant called mountain avens in the Rockies, jewelweed along a forest floodplain, and dandelions (to name four sets of ecotypes my students and I have worked on) all have locally adapted populations we call ecotypes. But ecotypes also occur in virtually every plant life form, from tall, long-lived trees such as Douglas firs, down to weedy annuals. The number of documented ecotypes is certainly in the thousands. To look for such differences and *not* find them has actually become the exception.

Plants form ecotypes in response to all manner of factors in the environment. Heavy metals at the mouth of a mine can select for differences in tolerance to those metals across a distance of a few feet. Others are adapted to other peculiarities of soil substrate, including pH (acidity), cation concentration, soil compaction, and even snow cover. And many species seem to be well adapted to the temperature regime where they are found.

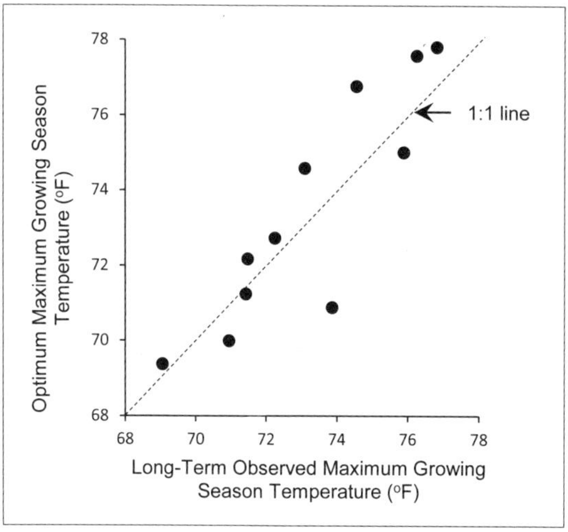

FIGURE 8.5. Relationship between long-term mean maximum growing season temperature average and optimum temperature for population growth in twelve populations

The bell-shaped curves we see in figure 8.4 suggest adaptation of ginseng to the local temperature regime, but the real teller is whether the peaks of those curves—the optimum temperatures—correspond to the temperature at the site where those populations are found. In one of the most satisfying graphs ever to come from my lab (fig. 8.5), we find the answer. There is almost a perfect one-to-one correspondence between the local temperature average and the optimum seen in the response curves.

The Hazard of Overspecializing

Adapting to a changing landscape is hard, especially if the change is very fast. Let's travel back in time. It is 1960. Your father's typewriter develops a glitch; the letter "a" is not printing correctly, even though the ribbon is new. Being mechanically minded and good in high school shop, you take the typewriter apart and find and then fix the problem with the mechanical linkage between the key and typebar. Everyone heaps praise on you for your technical prowess. After high school, you clearly see the business opportunity and open a typewriter repair shop. For twenty-five years, you operate a viable business. Then the technological environment shifts. Everyone begins to use computers with electronic keyboards instead of typewriters. Fixing those computers, keyboards, and printers requires an entirely different skill set. Soon your business is forced to close, and you are out of work. The environment within which you developed your business model has shifted. Many think the same kind of shift is happening now as we find ourselves at the beginning of a massive shift from gas to electric cars.

Becoming a specialist in your current environment, though, is a double-edged sword. You may succeed to a high degree, as long as the environment stays the same. Indeed, you could say that specialization is the key to success in an unchanging environment. Specializing in rapid and convenient oil changes can be a lucrative business (looking at you, Jiffy Lube), as long as internal combustion engine vehicles are the dominant mode of transport. But as we have seen from the examples above, when the environment shifts out from under you, you're endangered.

Fortunately, we humans have brains, and we can anticipate trends such as typewriters going by the wayside or internal combustion cars giving way to electric cars. That gives us a certain flexibility, and, if we're smart, we start retraining for new jobs when we see our old jobs going extinct. In our rapidly changing world, being adaptable is a good thing, a necessity almost, lest you become a dinosaur. It is no different in science. You either learn new techniques and methods, along with new ways of looking at problems, or you don't get funding, you don't get graduate students, and your research gets published, if at all, in backwater journals that no one reads. As a side note, many colleges and universities tout the very ability of their graduates to adapt to a changing world as the primary value of a "liberal education." Learning how to learn is more valuable than any particular set of facts you memorize. But I digress.

Now imagine a ginseng population, with a set of individuals all specializing in the current average temperature regime. But the climate begins to warm, and it does so more rapidly than at any time in the history of the population. Suddenly most years are on the right side of the curves in figure 8.4. The population is falling off its "adaptive peak" where it enjoyed success for centuries. And ginseng has no brain to anticipate such changes and learn their way to success in a new temperature regime.

Sadly, it's not just the southern populations that will fall off their adaptive peaks when the climate warms. The central and northern populations will fall off as well. Thus, no matter where you find yourself in the range, rapid climate warming is bad. By contrast, if the ecological model had been true for temperature response, ginseng would die out in the South, do worse in the formerly optimal center of the range, and do better in the North. At least somewhere it would do better, and the species as a whole might survive the cataclysm of warming. However, when each population is specialized, there is no region where ginseng is doing better in a warmer climate.

Because of the long-term data gathered on ginseng in many populations, this species has provided us one of the clearest pictures of the predicted response to climate change of any species on the planet. But recall, ginseng is not special. Most plant species are locally adapted. This means most native plant species, in turn, will likely respond like ginseng and do worse in a rapidly shifting climate.

What will happen in the long term if most of our native species of plants are going downhill because of a degrading climate? Ecological theory tells us that nature abhors a vacuum—in other words, some other species will fill the void left by loss of natives. Two groups seem likely candidates to succeed in the future climate. First, our "weedy" native plants that move around rapidly and succeed where other plants fail. In the eastern deciduous forest, this could include trees such as red maple and black locust, or understory plants such as nettles (in moist sites) and blackberries or greenbrier (in drier sites). The second group includes nonnative species from other countries that fall under the umbrella of invasive species. Tree of heaven (ailanthus), multiflora rose, Japanese barberry, Japanese honeysuckle, and kudzu are likely to spread. In the understory, garlic mustard and Japanese stiltgrass are just two of the hundreds of invasive species that could benefit. Much more on this important topic follows in chapter 11. In other words, in one hundred years, the eastern deciduous forest could look very different due to the high degree of specialization of our native plant species, resulting in their demise and replacement.

Climate Change Alters Harvest Effects

When you work so hard to collect field data for so many years, you want to make the most of it. And no one in my lab was better at squeezing interesting information out of dry data tables than Sara Souther. She realized that not only could we project population growth rates given annual variations in temperature we see now, we could also forecast what would happen if the frequency of warm years increases in the future. On top of this, we could overlay harvest in these model forecasts to see if climate change altered the effects of harvest. The short answer was that it did. And the pattern was simple. Harvest reduces population viability in the present environment, but harvest effects will be much more severe under a warmed climate. This was not good news for the long-term sustainability of ginseng harvest in the future.

Odd Climate Effects

In late March 2007, WVU students had been wearing shorts and T-shirts for two weeks already and were delighted to think that winter was over. Signs of spring were all around. Apple and peach orchards were in bloom. Roadside coltsfoot was in bloom. Daffodils were dying back already, and tulips were in full flower. Even some of the trees began to leaf out as the balmy temperatures provoked them to break dormancy. In the woods, spring beauties and trillium were coming up a full month before their usual season.

Ginseng is not an early riser in the spring. With its vegetative bud often an inch or two deep in the leaf litter, it remains insulated from the occasional warm midwinter day. But 2007 was different. The warm spell lasted long enough to heat the top layer of soil, and the stalks emerged from the litter, reaching up toward the canopy. Leaves were unfurling. Then the big freeze came.

In the first week of April, the atmosphere's jet stream warped southward, allowing cold Arctic air to dip south as well, yielding multiple days of subfreezing temperatures across a wide band in the eastern United States. Green plants were sitting ducks. Satellite images showed vast greening areas of forest and agricultural land turning brown. In North Carolina alone, crop damages were estimated at more than $100 million. Of our thirty censused ginseng populations, fourteen were affected by the frost event, with up to a third of the plants showing frost effects when we visited later that spring. Southern and western populations were more affected than northern and higher-elevation populations. The greater warming that occurred in the former more effectively stimulated early growth of the normally slow-emerging ginseng plants. Large plants were slightly more affected than small plants. Frost-damaged plants also grew more slowly the following year. Even reproduction was affected the following year, though mortality was not.

Climate models suggest that extended winter warming events are more likely in the future and that extremes will be more pronounced. The combination of these factors, warming followed by freezing, can trigger a maladaptive response of native species such as ginseng. Who would have thought that climate warming could result in greater freeze damage?

Climate can also disrupt mutualism. Typically, life cycles of pollinators and flowering plants are exquisitely timed to overlap, providing benefits to both. Climate change could alter that timing, and the evidence for that is showing up in several plant pollinator systems around the world. Unfortunately, no one has yet taken up this question with American ginseng. We do know that pollinators are crucial to ensuring some degree of cross-pollination among plants, reducing inbreeding. We simply don't know whether this mutualism is being disrupted by climate warming yet.

IMHO

To me, climate change is the defining issue facing humanity today. I am not talking about the defining environmental issue . . . the defining issue, period. I was being deliberately provocative by quoting sixteen-year old Greta Thunberg at the beginning of this chapter. Greta is actually articulating nothing new. She is simply presenting with a new voice the crystallized concerns about earth's future that scientists have been expressing for the past thirty years. But the way she frames her argument—about human suffering, about the tragedy of extinction, about the way powerful people are heedlessly jeopardizing her and her peers' future—in blunt, angry, tearful, and impassioned rhetoric—is a far cry from scientists' academic writing in peer-reviewed journals or even the Intergovernmental Panel on Climate Change report.

Greta is broadcasting her voice in the halls of power, from the European Union, to the United Nations, to the U.S. Congress. And people are listening. There are debates about whether scientists would be more effective adopting Greta's approach. After all, just because they have a PhD degree does not mean scientists are not people, with feelings, with children and grandchildren, and with deep attachments to a threatened natural world. It is not that they are not angry. They are. They want to maintain an air of credibility with reasoned discourse. And, truth be told, they fear the vitriolic backlash that flows on social media. Greta is fearless.

Humans are not leaping to do what needs to be done about climate change because we seem fundamentally incapable of taking our long-term future seriously. For some, the concerns of the now are simply too great to spend time worrying about the future. For others, fear of the unknown is paralyzing. Still others believe the science but think personal sacrifices in the here and now are too great. So, while we are as gods in our ability to impact the planet, we fall far short when it comes to acting in our own best interest to preserve it.

No doubt, ginseng is a bit player in the global climate change drama. The forecast about ginseng's response to climate change is bleak, for sure, but it is not why humanity will tackle the issue, if we choose to. Nevertheless, if you care deeply about this plant, you will care about whether society acts to mitigate climate change. In my opinion, of the three explanations for the decline of ginseng, climate change is the most serious issue in the long run and ultimately could be the cause of its demise.

CHAPTER 9

Appalachian Outlaws

Reality TV to me is the museum of social decay.—**GARY OLDMAN**

Reality TV

In 2014, the History Channel aired a show called *Appalachian Outlaws* set in the mountains of West Virginia and depicting a cast of greedy ginseng diggers and dealers. Not to be outdone, the National Geographic Channel filmed a show called *Smoky Mountain Money* with four teams of "sengers" setting out to strike it rich in the hills and hollers of western North Carolina. This was reality TV, patterned after *The Real Housewives of Orange Coun*ty or *Ice Road Truckers*, complete with untrained local actors, hyped drama, staged conflicts, and reinforced stereotypes. They tapped into the prurient fascination of viewers wanting a window into lives more exciting, more depraved, and crazier than their own. Both shows were bad television with bad acting, made worse by the faint guise of legitimacy provided by the "History" and "National Geographic" umbrellas. They inadvertently encouraged trespassing, stealing, and poor harvest practices in the real world. Fortunately, the shows did not last. But were they reflecting, at least in part, a new reality? Was the romantic ideal of Art and Ellie, ginseng stewards, the real fantasy?

Two Case Studies

In the spring, censusing our thirty ginseng populations required a group of three: one person following the phototrail and flagging plants, then labeling the flags with sticky notes showing the plant ID number, a second person hunched on the ground making measurements, and a third carefully recording data and double-checking the measurer to be sure they remembered all the important information. In the fall, we reduced the team size to two as we were measuring fewer variables, mainly focusing on berry and seed counts.

On August 22, 2008, Sara and I were flagging plants in a Kentucky nature preserve (Population 3) when suddenly we heard an animal pawing at the ground just out of sight over the hill by the streambed. We looked at each other, puzzled by what we had heard, but all I could figure was that a large deer was kicking at the rocks and making the noise. Intent on finishing the berry counts of this large population in one day, we went back to work. Twenty minutes later, two young, bearded men in dirty blue jeans and T-shirts strolled up the path by our population and stopped to chat about the hot weather. They asked where the entrance to the nature preserve was, and we pointed in that direction. They proceeded on down the trail, and we noticed a small digging tool strapped to a partly full knapsack as they walked away. We practically ran down the path to see what they had done.

Our ginseng population had been harvested almost under our noses. The most dense ginseng patch in all thirty of our populations was devastated. Wilting tops were strewn about on the soil surface. Fresh piles of dirt were scattered in all directions. Later we discovered they had also harvested a second very dense patch a few yards away. These patches each had about one hundred plants in an area the size of a kitchen table. Though we were immediately saddened and a bit upset by what had happened (we had been following these remarkable patches with great difficulty for years), we realized this also represented a tremendous opportunity to document a harvest by actual harvesters. Also, perhaps foolishly, we hastily realized that it could represent an unprecedented chance to quiz the harvesters and learn about them firsthand. We sat down and scribbled a list of questions. First we thought it would be wise to tell them we had no intention of turning them in to the authorities but that we were scientists, wanting to understand the way harvesting was done. Had they been to this nature preserve before? How long ago? Would they plan to come back? When? Had they observed the numbered nails as they harvested, and, if so, what did they think of those? How did they decide which plants to harvest? How much time did they spend harvesting every year? We then ran after the harvesters to conduct the interview. They had vanished.

Believe it or not, these two diggers showed some of the most ethical behavior we observed in all harvest events we documented. The harvest was legal in terms of the date, since Kentucky's wild harvest season began on August 15 at that time (since changed to September 1). The harvest was illegal because it occurred in a nature preserve where harvest is explicitly prohibited. Based on the tops the diggers had discarded, we knew that 15 percent of the plants taken

FIGURE 9.1. Relative population size over a decade for two populations affected by illegal harvest in 2005–2008 (Population 6) and 2008 (Population 3)

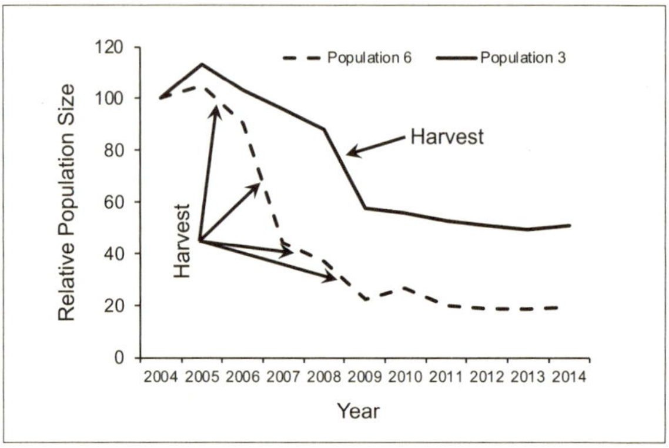

were one-leaf or two-leaf plants, though Kentucky law specifies only plants with three or more leaves could be harvested. The patches were so dense, however, that it would have been difficult to harvest only the large plants without unintentionally damaging neighboring small plants. The effect of this harvest on numbers was clear (Population 3; fig. 9.1).

A much worse harvest occurred in another Kentucky nature preserve, not too far from the first. Population 6 was a beautiful population with many large four-prong plants. It was one of a handful of our thirty populations we would describe as mature, in that it had a nice mix of plants of all size classes, along with the high proportion of very large seed-producing plants. My colleague Dr. Anne Lubbers of Centre College had been censusing this population with her students as part of our National Science Foundation (NSF) study for several years. In late July 2006, she was puzzled to find a large number of missing roots that had been there only weeks before in the spring census. Eventually, she concluded it simply had to be harvest. And it was a bad one. In this case, the harvest was illegal in three respects: location, removal of undersized plants, and out of season. Almost a quarter of the plants taken were undersized. However, the loss of seed production was most pronounced, because of the removal of a large number of monster plants. After hearing what Anne described, I remember trying to console her by saying that this was a really important event she had witnessed—that the human-ginseng interaction was exactly what we needed to know about, tragic as it might be. I don't think my words helped much. And unfortunately, digging continued annually in this population until it was a shell of its former self several years later, consisting of mostly small nonreproductive plants. In Populations 3 and 6, harvest was clearly the factor that most impacted population sizes (fig. 9.1).

Experimental Harvest

The harvests of Populations 3 and 6 were messy. I have no doubt that the two diggers in Population 3 would have taken more plants if we had not been discovered at the site during their visit. The Population 6 harvest was repeated annually, which is quite unusual, so the effect of a single isolated harvest event was not clear. Early on, Martha Van der Voort, Brent Bailey and I took advantage of an opportunity to carry out an experiment when a mountain biking path veered toward a local population we knew about. Concerned that it would be damaged, and also needing roots for another experiment, we decided to conduct an experimental harvest. With the latest census in 2019, this study has now run twenty-four years (see fig. 9.2).

If we were hoping for a clear story by doing an experiment instead of an observational study, in this case we were bound to be disappointed. For the first twelve years, the population appeared to be recovering, with a flush of seedlings in the two years following harvest, followed by gradual growth of those seedlings to adult size. The increase in adults, however, came to a halt in 2008 when a large oak tree fell across the population. A few individuals may have been buried under the log, but more succumbed later, possibly from fungi associated with the decaying log. Nevertheless, those three-leaf plants had begun to produce seeds prior to the treefall, and a second spate of recruitment of new seedlings occurred in 2009 and 2010. After loss of the three-leaf plants, however, the seed bank was depleted, and the numbers of seedlings slowly declined. Some died and some graduated to the juvenile two-leaf class. What is evident is that, while overall numbers have fluctuated wildly since the harvest in 1996, often exceeding the initial N, the size structure (the set of numbers in each size class) has not recovered after nearly a quarter of a century. If you observe the three-leaf plants in particular, prior to harvest there were eighteen,

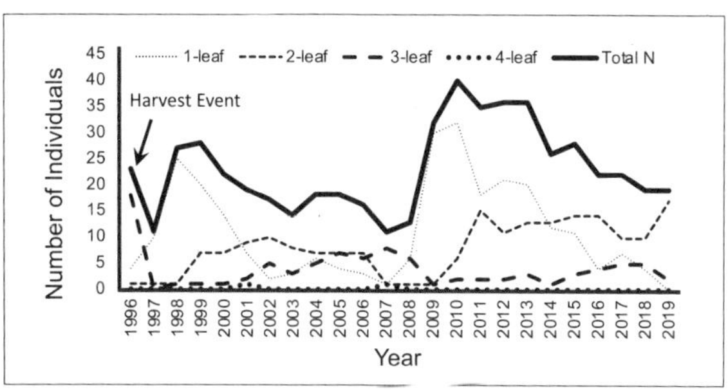

FIGURE 9.2. Recovery from harvest in a single small population in northern West Virginia

but the highest this number ever reached in the ensuing twenty-three years was eight in 2007. The bottom line is that a single harvest event had effects that could still be seen in the skewed size structure of the population twenty-four years later. Recovery, if it occurs at all, happens at a glacial pace.

How Common Are Stewards?

It would be easy to cherry-pick particular harvest events to make diggers look bad (e.g., case two above) or good, but that would not answer the fundamental question of how harvesters generally behave. In particular, it would not answer the question left by Martha's research (chapter 6) about how many harvesters fall into the stewardship, "compliant," or "noncompliant" (illegal) categories. After many years of censusing, however, we had accumulated enough observations of harvest (thirty-four) with our twice-a-year censusing that we could address this question head on. We divided the types of illegal behavior into three categories: place, time, and intensity. With respect to place, an illegal harvest was one that occurred on land where harvesting was not permitted, either by the private landowner or the managing agency of the forest. With respect to time, since the fall census occurred just prior to harvest season, we defined out-of-season harvest as any harvest observed between the spring and fall census. Since the fall census was not on the exact date of the beginning of harvest season, and because we assumed any census between fall and spring was legal, our numbers underestimate violations of harvest season laws. Finally, with respect to intensity, we examined the size distribution of harvested plants, noting when diggers removed plants smaller than the three-leaf threshold.

In our study, which spanned between five and eleven years (depending on the population), 43 percent of the populations were harvested during that interval. In the 221 "population years" of data, thirty-four harvest events were observed, suggesting an annual rate of about 15 percent of populations being harvested each year. The majority of harvest events (65 percent) occurred on land where harvest was specifically prohibited. About one-fifth (21 percent) of the harvest events occurred out of season. Finally, in 82 percent of harvest events, plants below the legal size threshold were removed. Overall, only 5.9 percent of the harvest events were legal in all three respects. These came the closest to "stewardship" harvests, though really, they were just compliant. Flipping that, 94.1 percent were illegal in at least one way. In other words, we were discouraged to learn that stewards, like our romantic ideal, harvesters Arthur and Ellie, are rare.

It would not matter that stewards are rare if most populations were never harvested, but the fact that nearly half our populations experienced harvest suggests that harvester behavior does matter at a broader scale.

What Drives Illegal Harvest?

It would be too easy to conclude that poor harvest practices are driven by simple greed. One could view ginseng plants as dollar bills scattered across the eastern deciduous forest just waiting for someone to pull them from the ground. Indeed, at the prices of a decade ago (ca. $300/pound and 250 roots/pound), each dried root was worth just over a dollar. Given that the vast majority of diggers just sell their roots to a dealer for cash, we can draw some conclusions by thinking of ginseng precisely in that way—as "found money." So what would drive a digger to take money from a place where it is illegal? Or pick a plant that is too small (worth only a half-dollar)? Or take the money in a way that prevents the place from yielding more money in the future (out of season; not planting seeds)?

Put this way, we can imagine many reasons for poor harvest practices. For example, if one views the government as inherently evil, or perhaps simply too intrusive in your personal life, for whatever reason, you might just rationalize that taking ginseng from a government-managed park or forest is just payback. For those who place a high value on personal freedom, government regulations may be viewed with suspicion or outright contempt, as laws are an attempt to constrain that freedom—in this case, freedom to find and collect that money that grows from the ground.

If you are unemployed, you might take advantage of the opportunity to harvest ginseng because you can sell it to get money to buy food to survive, and maybe it is one of very few options open to you where you live. Analyses have shown a positive relationship between harvest rate and unemployment rates. Or you might be desperate with a different kind of hunger. You might be addicted to alcohol or painkillers and need money to slake that thirst. You may have even found a dealer who will take ginseng in payment (this has happened). Could the ongoing opioid epidemic be shifting the harvest landscape away from stewards toward more noncompliant harvester types?

Taking plants out of season can be motivated by concern that can be expressed as follows: "If I don't dig it first, someone else will." In other words, the perception of competition for the increasingly scarce resource rewards early diggers, which simultaneously ensures that the resource will not be there in the future. By the same token, a digger might take plants they know are too small, because someone else will if they don't. This is the famous "tragedy of the commons," whereby it behooves each individual in the short term to take more than their share of resources from some common source. But if everyone does the same, the overall resource suffers, and everyone gets less in the long run. The typical example used to illustrate this principle is placement of personally owned cows onto a "renewable" commonly owned pasture. The pasture will be degraded by having too many cows per acre. But the principle also applies to removal of renewable resources by individuals from a forest that is viewed as a commons (even when it is private property, in fact).

Another reason harvesters might do the wrong thing is that they simply don't know they are harming the plant by what they do. There are numerous misconceptions about ginseng that have been propagated due to lack of scientific knowledge. The most serious misunderstandings concern how slowly the plant actually grows and how few seeds are produced. Ginseng is not a weed or a garden plant or a typical crop. Stewards know this.

Perhaps a harvester does not mind trespassing to dig plants because they are "just plants," and most landowners are assumed to not care. Certainly some court magistrates view ginseng this way when they hand out minimal fines or jail time for ginseng theft from private property. Like found money, ginseng roots are small and can be easily hidden in a knapsack, so no one will necessarily know what you have been doing unless you are caught in the act of digging. No shots have been fired. No large carcass has to be carried out. Enforcement is difficult.

Indeed, part of the calculus of any would-be thief is both the chance that they would get caught combined with the severity of the punishment if they are caught. In the case of ginseng, unfortunately the chance of getting caught is usually low. For example, in our system of national forests, there is one law enforcement officer per 262,000 acres under management, and their job entails enforcement of all kinds of laws, not just those governing plant poaching. The ratio may be slightly better in state parks and national parks, but again the majority of the ranger work involves people management in areas where visitors are concentrated, not in the hinterlands of the parks. Private landowners may have the best chance of catching a ginseng thief, but, once caught, a judge is likely to fine them a few hundred dollars at best, with perhaps a very short-term imprisonment.

Illegal harvest is motivated by short-term economic gain, but it comes at the expense of the long-term welfare of the plant and ultimately the long-term economic welfare of harvesters. Illegal harvest is shortsighted. But then, how many human failings are due to shortsightedness? How many of the most serious problems we face could be solved by taking the long view?

Zooming Out

This chapter has once again zoomed in on harvest. The data presented here, accumulated by studying dozens of wild populations over decades, have sharply altered our initial "romantic ideal" view of harvest. The sad truth is that harvest practices seem far from the Art and Ellie stewardship narrative. Perhaps long ago that was the norm. But, more likely, humans have not fundamentally changed. There were likely always a few ginseng "stewards," as there are today, and a larger group willing to follow the rules, if they existed, but not go the extra mile to be true stewards. But a substantial fraction of diggers are actual outlaws. Not the kind of outlaw to be romanticized in novels or cheesy reality TV shows but lawbreakers—trespassers and thieves who threaten ginseng with long-term harm. Any attempt to conserve wild ginseng must confront this reality.

Yet, even with all this additional discouraging information, we were still not sure that harvest was ginseng's worst problem. Deer browse, climate change, harvest. Three strikes. But this isn't baseball. What else could be limiting ginseng's return to the glory years of the presettlement old-growth forest habitat ginseng seems to prefer? Clues were written on the landscape, if only we could read them. We needed to zoom our scientific lens back outward to place the effects of harvest in a broader context before assessing its conservation impact.

IMHO

Wishing does not make it so. Much as I wanted to believe the prevailing academic myth of the benevolent, wildcrafting, stewarding ginseng harvester who takes the long view and has ginseng's welfare first in mind, the data on wild harvest largely contradict this view. The most protected public lands, our national parks, struggle to combat poaching every year. Even Park Service rangers know they catch only a fraction of poachers and that law enforcement will likely exact a weak punishment if the perpetrator is taken to court. Parks are working to change this, to make the penalties more severe to compensate for the time and effort required for restoration. But if these crown jewels of our public lands are struggling, imagine how much more difficult the problem is for state and county parks, let alone national and state forests. Changes to existing protections for ginseng are needed, and when they are devised they need to account for the actual behavior of harvesters, not an assumed fantasy. This means implementing enforceable regulations in which compliance can be checked and opportunities for cheating are minimized.

CHAPTER 10

The Ever-Changing Landscape

> One of the penalties of an ecological education is that one lives alone in a world of wounds. Much of the damage inflicted on land is quite invisible to laymen. An ecologist must either harden his shell and make believe that the consequences are none of his business, or he must be the doctor who sees the marks of death in a community that believes itself well and does not want to be told otherwise.—**ALDO LEOPOLD**

If there is a choice, I always take the window seat. From an altitude of thirty-five thousand feet, the landscape reveals patterns, at once beautiful and illuminating. Traveling west from the East Coast, the megalopolis-covered coastal plain quickly gives way after takeoff to forest-covered foothills. Irregular patches in the forest tell the history of timber harvesting. Squinting, you can perceive the grain of the forest—fine grain indicating young dense stands, coarse grain indicating older, maturing stands. Traversing the Appalachians, the long northeast–southwest ridges are punctuated by grassy, cleared valleys. The ridge-and-valley section is brief, and the mountains soon become dissected by streams flowing in all directions. This landscape is heavily forested, and cleared farmlands are the rarity. Gradually, mountains give way to hills, and clearing becomes more frequent. An odd form of clearing appears in southern West Virginia and eastern Kentucky. Whole hilltops have been sliced off and hauled away and the debris dumped into nearby streams, in a testament to human power—mountaintop-removal mining. Old strip mines are visible as grassland and sparse forests that wind around the contours of the landscape. Moving farther west as the contours flatten, farms become larger. Pasture, along with fields of corn and soybean dominate the ecosystem, and forested parks become the rare landscape patch. Finally, headed toward Los Angeles or Denver or Seattle, you traverse the vast, grassy, treeless plains.

The entire landscape you have passed over is a snapshot; it has changed radically over time. What does all this mean for a shade-loving understory plant? As it turns out, quite a lot. In this chapter, I zoom out in space and zoom back in time. Our previous close focus on what happens in the present to individual plants (they get browsed, they get harvested) or populations (they decline or they increase) is only part of the story of ginseng. Sometimes these up-close observations compel us to expand our perspectives to comprehend the larger scale forces at work. Some of them make us view the landscape from above with the different sensibility of an understory plant.

Mature Forest Lover

It was the most beautiful Appalachian cove forest community I had ever seen. Unlike other treasured spots, such as the Joyce Kilmer Memorial Forest in North Carolina or the Core Arboretum in Morgantown, West Virginia, it was not a true virgin forest or even an old "old growth" forest, since it had been cut somewhere between 1903 and 1911. Nevertheless, over the century that followed, evidence of past disturbance at the site had all but disappeared. The overstory canopy consisted of a diverse set of trees of wide-ranging sizes. The largest of them were three to five feet in diameter. Majestic red oak, black cherry, tulip poplar, sugar maple, basswood, and beech were spaced across the gently sloping site, with a generous mix of smaller trees scattered beneath. Decades of leaf litterfall had produced a rich organic soil layer. The understory was even more diverse than the overstory, with maidenhair fern, purple-fringed orchids, black cohosh, nettles, and jack-in-the-pulpit, among a few dozen others, and virtually no invasives that are common in more disturbed or more heavily browsed sites. Located above twenty-five hundred feet, its temperatures remained relatively cool, even in midsummer, despite the eastern, southeastern, and southern aspects of much of the site. The browse rate of ginseng plants was the third-lowest of our thirty populations. Exactly zero plants had been harvested in the twenty seasons we visited. The population had grown 69 percent over the monitoring decade (2004–2014), an average of 5 percent per year, the greatest of any of our populations (see the top curve in fig. 4.5).

I often refer to Population 30 as the "quintessential ginseng population," or the "best population." It certainly was the best in terms of increasing the most and in consisting of a healthy mix of young and old, small and large plants. But there was more to this place. My whole field crew loved visiting P30 every year, marking on our calendar the days until we would be forest bathing in a landscape where we did not feel like we were living alone in Aldo Leopold's world of wounds. This was a place where the flutelike song of the wood thrush, the "see see see Su-Zee" of the black-throated green warbler, and the high-pitched trills of the winter wren could be counted on to provide an enchanting avian soundscape to accompany our note-taking.

But why was Population 30 doing so well? Lack of harvest, low browse rates, and high elevation with less heat and drought were no doubt part of the answer, but we had other populations with those traits. A clue came from an off-the-cuff observation made while hunched over plants at this spot year after year: we noticed that we were often in the sun. Lying on your back looking up at the canopy, you could see why. Despite the high treetops, there were scattered holes in the canopy created by medium to large trees that had toppled. Ecologists refer to these holes as canopy gaps. Over many years, surrounding trees would fill in these gaps from the sides or by growth from below, but in the short term, when an overstory tree fell, an infusion of light reached the forest floor, bathing the understory plants for anything from a few minutes to two hours at a time. Responding to this burst of energy, ginseng and other understory plants would rapidly increase their photosynthetic rates, assimilating more carbon into sugars and starches to grow larger roots and feed their energy-requiring flowers, fruits, and seeds.

Younger forests lack these canopy gaps. In an old field or a clear-cut, the highest densities of trees occur just a few years after abandonment or cutting as seeds sprout in abundance and, in the case of cutting, stump sprouts gush forth new stems. Then the competitive war of attrition begins, with the smaller, weaker trees falling out as rapidly growing early successional trees crowd each other, reaching skyward, usurping light from that equally light-seeking neighbor. The losers in the vertical race for sunlight die from lack of energy to maintain basic functions as they get shaded from above, and this density-dependent thinning continues for decades. Fewer and fewer trees remain per unit area, but they are larger on average. All the while, herbaceous understory plants, wholly lacking any chance of competing with trees for light, are heavily shaded by the dense canopy above. Only when the canopy trees reach large sizes and older ages does the death of a canopy tree leave a substantial gap, allowing patches of sunlight to reach the forest floor. The patches are transient, moving generally west to east as the sun arcs across the sky east to west. Understory plants sit in these light patches for short periods. Those that can ramp up their photosynthetic rates in response will add to their energy storehouses much faster, all the while avoiding the drought stress imposed by long hours of direct sun.

At this point, the canopy gap hypothesis was just that: a reasonable guess based on known physiological responses of plants. It took careful work by a dedicated undergraduate student, Alixandra Wagner, to gather the scientific evidence, and there was no better place to do this than Population 30. Alix was one of our most reliable conservation interns, always careful, helpful, and quietly competent at everything she did. I was lucky once again to have found a multitalented student, one who was not only brawny enough to lug a forty-pound "portable" Li-Cor photosynthesis system to the site but also bright enough to learn to operate it successfully. Photosynthesis and transpiration are measured with an infrared gas analysis (IRGA) system. This instrument, which costs about the same as a new Dodge Ram pickup, measures the rate of disappearance of carbon dioxide from an airstream passing over the leaf inside a chamber. Simultaneously, it measures the increase in humidity caused by water loss from the leaf's surface. When I was an undergraduate, these systems occupied the better part of a lab the size of your kitchen. As with the evolution of computers, IRGAs shrank over the years by technological evolution to fit into a large suitcase, but they were still finicky machines that required a basic understanding of plant physiology and complex electronics. Students who went down the rabbit hole of Li-Cor IRGA technology to become superproficient were dubbed IRGA-nauts. Alix was not exactly an IRGA-naut, but she learned enough to get reliable numbers, and that was enough to test the gap hypothesis.

Alix discovered that ginseng did in fact respond to intense sun with dramatic increases in photosynthesis. Not only that, plants that were in micro-sites with more exposure to sun patches built leaves that were better able to take advantage of the high light when it came. This translated into faster between-year growth for plants in the better-lit areas of the population. In fact, at micro-sites with the lowest exposure to canopy gaps, plants were shrinking in size, on average, while those with higher exposure to these sun patches got bigger. The only downside to being in a high light micro-site was that seed germination was reduced. We hypothesized from this observational study that ginseng was well suited to take advantage of light gaps that form in mature forests, whether or not they were true old-growth forests. As with all observational studies, it is dangerous to draw a firm conclusion without a more rigorous experimental test, but the hypothesis seemed a reasonable one. At that point, we weren't ready to begin chain-sawing down large trees to test the hypothesis, especially not in our favorite population of all.

Timber Cutting and Ginseng

Though Population 30 was sacred, PhD student Jennifer Chandler realized that tree-thinning experiments were going on all the time in forests across Appalachia. It was called timber cutting, and there were lots of different kinds of tree harvesting methods being practiced, from selective tree harvests (e.g., eighteen-inch diameter minimum) to patch cutting. Could harvests that cre-

FIGURE 10.1.
Burst of fruit production from a large plant that survived a heavy timber cut in Population 18 in Virginia. At the same site, over several years, most of the plants present before the cut died.

ate canopy gaps like those in old-growth forests actually help ginseng in the understory? The trick was to find timber-cutting operations overlapping with ginseng populations, preferably a year or two before the tree cutting began so a baseline could be established. With a lot of detective work, Jen was able to find five areas with both ginseng and timber cutting to gauge the impact of these practices.

As we often found with this species, ginseng's response to harvest of overstory trees was more complicated than we initially thought. Survival of plants suddenly exposed to the sky by timber harvesting was generally lower compared to what it was before the cuts; in one intense cut more than half the plants died the first year. Some of the mortality was caused by soil disturbance owing to the timbering operation itself, but in other cases death appeared to be a result of drought stress. By contrast, the growth rate of plants that did survive increased after cutting. This observation was consistent with stories we heard from ginseng diggers claiming that the largest plants they ever found were in clear-cuts. Plants in moist micro-sites that survived the canopy opening could become huge within two to three years (see fig. 10.1). There was also evidence for increased efficiency of seed production. Decreased seed germination was expected to result from the greater desiccation of seeds at the soil surface. Jen and I concluded that ginseng was remarkably resilient in response to disruption of the canopy caused by timber cutting but less resilient to soil disturbance caused by logging roads and dragging of logs out of the forest. Overall, we concluded that the more intense the tree harvest, the greater the ginseng population size reduction would be.

Jen followed up her study of five tree-harvesting sites with a detailed look at the physiological responses of ginseng in the year prior to tree cutting, as well as one and two years after. Like Alix, she lugged the Li-Cor to her field site on many occasions and learned the intricacies of IRGAnautics. Her work largely reinforced what Alix had shown before in Population 30 with natural variation of light levels. The enhanced exposure to sky after tree cutting was a stimulus to photosynthesis and growth. She did show, however, that full photosynthetic acclimation occurred only in the second year following the cutting. That extra time was needed for the plant to construct the thick leaves with more chlorophyll that were better able to take advantage of the increased light.

Natural Canopy Openings

Further building on Alix's undergraduate work, Jen's PhD led her to extend the canopy-opening studies to instances where the canopy was opened suddenly by natural disturbances. In our thirty populations of ginseng, the long-term censusing really paid off here, because by chance we encountered three major natural canopy-opening events at three different populations: (1) an ice storm that ripped out large branches and felled whole trees, (2) a forest tent caterpillar outbreak that defoliated the trees for several weeks one summer, and (3) a lightning strike, which killed multiple trees. Of course, none of these events was an experiment that could be replicated, but together they provided a window into the world of understory plants and the factors that influence their lives.

In all three natural canopy-opening types, the effect on population growth was positive, suggesting a stimulating effect of additional light reaching the forest understory consistent with all the previous findings. When Jen broke down the cause for this difference, it was always related to greater growth rates, which in turn led to larger plants over time, that in turn produced more seeds.

When ginseng diggers get together, it is usually not long before the conversation turns to the size of the biggest ginseng plant they have ever seen, much as fishermen compare their trophy catch. In our thirty natural populations, I have observed monster five-prong plants only a handful of times, and each time it

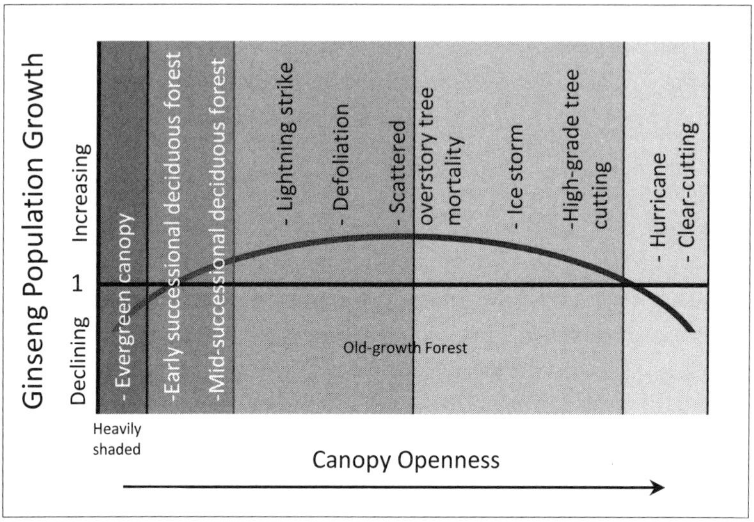

FIGURE 10.2.
Response of wild American ginseng to canopy openness, with heavily shaded overstory on the left and massively open canopies to the right

was after a canopy opening. The plant was already a large and robust three- or four-pronger, and the extra light gave the plant the energy boost needed to become extra big the following year.

Model of Ginseng's Sun-Loving, Shade-Loving Contradiction

Ginseng growers use shade cloth that transmits 20–30 percent of sunlight to get maximum ginseng growth. As you walk through a fifty-year-old deciduous forest, you might think the canopy is equivalent to shade cloth. But our eyes deceive us. Our pupils open wide to compensate for low light conditions. In a closed-canopy deciduous forest, more typical light levels are 5–10 percent of ambient, but they can be as low as 2 percent. All of the above experiments led us to an overall parabola-shaped model of ginseng's response to canopy opening (see fig. 10.2). Population 30 would land squarely in the middle of the curve.

In the Context of History

By 1880, although the lowlands of New England, the Eastern Seaboard, and even much of the Midwest had been cleared for farmland, most of Appalachia was still covered with ancient forests disturbed by occasional fire, flood, or hurricane but largely intact. With the extension of narrow-gauge railways into the mountains, along with invention of band mills that could process huge amounts of timber, over the next forty years virtually all of that old-growth forest was cut. Had there been airplanes all that time, the view from them would have undergone a rapid shift. By 1920, truly virgin forests that had been spared the saw were extremely rare. One estimate is that today we have 0.1 percent of the original forest left. I suspect it is less. Jennifer Chandler's research showed how ginseng could survive such a cataclysm. Populations were no doubt diminished in size, but the occasional large plant that survived left a lot of seeds, some of which would successfully propagate the population.

Perhaps even more important than the initial stress of timber cutting, however, was the shade squeeze imposed on understory plants, the intense shadows cast by early successional forest canopies that followed the cutting. In Appalachia, populations survived from the 1920s through midcentury in this suboptimal habitat. The dramatic reduction in ginseng harvest figures that were seen relative to those in the nineteenth century could be due, at least in part, to the massive expansion of logging in the late 1800s and early 1900s, followed by decades of heavy forest shading.

Are Forests Getting Better for Understory Plants?

Although some areas of regenerated forest have experienced a second wave of timber harvest, the next wave of tree cutting has been more drawn out in time, more variable in intensity, and often has not occurred at all. At least not yet. Across the landscape from New England to Georgia, west to Missouri, and north to Minnesota, more forests are maturing to the point where large overstory trees are beginning to fall out naturally, forming canopy gaps that allow life-giving light to reach the forest floor. More of them look like that coarse mosaic as seen from the air. Owners of private land have begun managing their forests for different purposes, not just to sell wood. Some land managers have begun to rethink the purpose of the land they manage. Is the forest there to produce successive crops of wood? Or is there more to ecosystem health than board feet of lumber and dollars they bring?

Over much of the eastern deciduous forest, deer overpopulation is the main brake on recovery of the forest understory. Plants naturally tend to increase until competition takes over and they become density-limited. But too many hungry mouths that will eat almost anything green to stay alive can turn the understory into a veritable desert.

IMHO

To sell their services, foresters often use the term "overmature" to describe the condition of a given private woodland. To me as an ecologist, this term would be merely laughable if it were not also deceptive and misleading. It implies something is wrong with the forest and needs human intervention to correct the problem. While it is perhaps true that the forest is not adding marketable wood at the rate it was previously, on a per acre basis, in almost every other sense that forest is becoming more valuable as it matures toward old-growth state. First, mature forests have tremendous value for wildlife. Large holes in huge trees provide nesting spots for everything from owls to wood ducks to flying squirrels, not to mention denning sites for black bears. Second, as we have seen, the scattered holes in the canopy of a mature forest stimulate understory plant growth, which in turn feeds pollinators with nectar and nourishes wildlife with leaves, fruits, and seeds. Mature forests are even valuable in the global carbon budget, storing more carbon than their early successional counterparts in the deepening soil, in decaying logs on the forest floor and in wood and leaves in the canopy. We need more mature forests, not fewer, for conservation of plants and animals and ecosystem health. Kudos to landowners and public forest land managers who recognize the value of mature forests in the larger disturbed landscape. I realize that not all landowners can afford to let their forests age in place. But let's get innovative and assist them in doing so because of the great service they provide for biodiversity preservation. Let's consider paying landowners not to cut special mature forests, much as we already compensate farmers for retaining wetlands because of their wildlife and flood protection value. Why shouldn't retention of old, mature, diverse forests be compensated, at least until they reoccupy a significant fraction of the landscape of the eastern United States? One-tenth of 1 percent is not enough.

Semipermanent and Permanent Landscape Changes

Logging was not the only impact European settlers imposed as they occupied the eastern deciduous forest landscape. Forests were lost to settlement of towns that became cities that became multicity megalopolises. The easiest way to comprehend the scope of this change today is to zoom out beyond airplane altitudes to space and from there to observe nighttime light emissions (fig. 10.3).

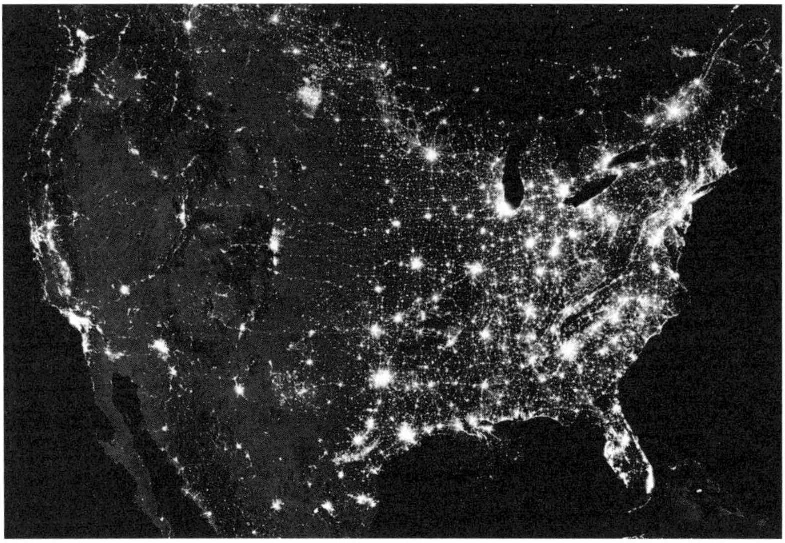

FIGURE 10.3.
View of the continental United States from NASA-NOAA Suomi NPP satellite imagery in 2012. Urban and dense suburban areas emitting such concentrated light are largely lost as potential habitat for species such as ginseng. (Image from NASA.gov)

Urban and suburban sprawl. Despite the large concentration of nighttime light in the eastern United States, the urbanized land area is still relatively small. Connecticut, Massachusetts, New Jersey, and Rhode Island are all between 30 percent and 40 percent urbanized. Believe it or not, New York is only 8.7 percent urbanized. The central ginseng states are all less than 10 percent urbanized as well (North Carolina, 9.5 percent; Tennessee, 7.0 percent; Virginia, 6.7 percent; Kentucky, 3.6 percent; West Virginia, 2.7 percent). Though urbanization is not the main cause of habitat loss, in the big scheme of things it is significant. As the human population grows, this loss has been steadily taking ground and is likely to be permanent. Although it is not unheard of, few towns have disappeared back into the forest over the past century.

We have to be cautious about excluding all suburban areas as ginseng habitat. If housing development involves felling all or most of the trees, bulldozing the land, planting grass, and then erecting houses, then ginseng habitat is gone. But often, especially in recent years, homeowners are valuing the privacy offered by intact woods as well as the lower maintenance required by smaller lawns. Developers are therefore more often leaving second-growth forests intact surrounding new homes. Sometimes these woods contain a complement of native understory plants. The largest challenge in such areas again turns out to be deer, since hunting is often not allowed near homes in the woods. Invasive plant species can be another problem in these areas, particularly when homeowners become complicit by planting species such as Japanese barberry in their landscaping.

In the development where I live, seventeen homes are located in the woods on about forty-five acres. Covenants require property owners to get permission to cut any tree over six inches in diameter. Most homeowners prefer their lots wooded and unfenced to allow free movement of wildlife and to reduce the burden of lawn care. The homeowner group recognized the deer overpopulation problem and cultivated a select group of bowhunters who cull the herd each year in the surrounding nature preserve (and sometimes in backyards of property owners). Jack-in-the-pulpit, mayapple, spring beauty, cutleaf toothwort, orchids, geraniums, and a wide variety of other native understory plants grow abundantly under a diverse overstory in the woods around the houses. Ginseng would grow here, though invasive plants are an issue, with garlic mustard, Japanese stiltgrass, autumn olive, multiflora rose, tree of heaven, and oriental bittersweet spreading in certain areas.

Among the thirty ginseng populations in our study, only one (Population 12) was located in such a place, in this case a suburban Pennsylvania forest adjacent to a lot where a house was built during our study. We examined the effect of the lawn edge that was created during our census period. The main edge was on the north side of the forest fragment, and therefore enhancement of light penetration into the ginseng population was minimal. Indeed, my master's student Zach Bradford showed that "edge effects" on ginseng performance there were minimal and that ginseng continued to grow in the fragment, though numbers declined by nearly 50 percent over the monitoring decade. At three other forest margins, Zach showed a reduction in ginseng pollinators closer to the edge. He also showed that pollinator visitation appeared to be related to seed production.

Deer, invasive species, and edge effects all diminish the suitability of suburban forests for ginseng, but these places are not always uninhabitable. Indeed, the proximity of forest-valuing humans to these populations opens the opportunity for stewardship to help them grow. Suburban forests should not be written off as biodiversity preserves.

Forest to farm (and back). Zooming back in time again, the largest cataclysm understory plants experienced immediately after European colonization of North America was wholesale conversion of the flatter portions of the landscape to farmland. From the 1600s forward, this great agricultural clearing process spread like a westward-flowing wave across the continent. Colonists who traversed the Appalachians to find arable soil in Ohio, central and western Kentucky, and Tennessee quickly cleared those lands, planting crops suitable to the climate. Onward the wave spread to Missouri, Indiana, Illinois, Michigan, Wisconsin, and Minnesota. All ginseng country. After forest clearing, fields were tilled and crops planted. Even relatively hilly portions of Appalachia were cleared, with widespread pasture for sheep and cows, as well as hayfields and orchards. Woodlots were retained on many farms, but they were islands surrounded by fields and typically located on the least fertile soils. Only the rugged mountains and steep ravines or gorges escaped the wave of agricultural land conversion. The peak extent of farmland dominance of the landscape differed among regions, but what many today do not realize is that there was a "forest transition," a point in time when the gain in forest land through abandonment of farms exceeded the rate of forest clearing. The transition occurred at different times in different areas of the country. For New England, farm abandonment began in the mid-1800s, while in Appalachia this process started in earnest in the early 1900s. Only recently has the gain in forest land area been stemmed by factors such as urban and suburban sprawl, second home development, and mountaintop-removal coal mining.

Before moving to the part of the story where I assert that reforestation has beneficially reclaimed potential habitat for ginseng and other understory plants, I think it is useful to try to capture what was lost as family farms declined, at a very human level. This may seem like a tangent, but please bear with me.

Change Measured in a Lifetime

As I write this in November 2019, I am on a vigil in a hospital. My father is dying. He lived a long and full life until his own cells betrayed him. I am once again in the window seat, looking at his life from aloft. James C. McGraw was born in 1928 on a small family farm on the edge of Appalachia in eastern Ohio. That was about the time of the forest transition, when land in the region was more cleared of forest than at any time before or since. His grandfather's farm on a hilltop outside Blaine was known as Highland Fruit Farm. They grew twenty choice varieties of apples, three varieties of pears, three varieties of grapes, and two varieties each of peaches, cherries, plums, quince, and gooseberries. Also on the farm were large vegetable gardens, chickens, milk cows (Blackie, Whitey, and Jersey), and workhorses named Bessie and Luke.

With rare energy in my father's voice, he tells me all these details now as if it was yesterday.

The apples and pears and peaches and eggs and milk my father remembered to the end of his life always tasted better than any that could be found at a grocery store today. In fact, throughout his adult life, my father reminisced about his youth on the farm in rapturous tones. He was a perfectionist for whom most things in life could not measure up, but "the farm" had a working harmony about it that embodied perfection to him. I am sure this romanticized notion of rural paradise, distorted somewhat by time, did not adequately account for the grueling day-in and day-out work of the small farmer's life, a life that his own father rejected. But thousands upon thousands of such beloved enterprises were scattered across the landscape of the eastern United States and were central to the economy of a nation.

The migration of folks from the family farm to urban areas for jobs was part of a diaspora that was not just an Appalachian phenomenon. It was a regional and ultimately a worldwide one. It has been well documented by human demographers but better captured in spirit by musicians the world over, perhaps none more poignantly than musicians who grew up in ginseng country. Jad Abumrad, host of the popular podcast *Radiolab*, grew up in Tennessee and recently returned there to try to understand the global appeal of country singer-songwriter Dolly Parton. In a series of ten podcasts (DollyPartonsAmerica.com), he traces at least part of her broad appeal to her nostalgic portrayal of growing up on a small mountain farm in East Tennessee. And, of course, she wrote a famous song capturing the peaceful life she knew there, "My Tennessee Mountain Home."

With her lyrics, Dolly Parton taps into that sense of connection my dad felt so deeply, along with millions of others who felt at their core the loss from leaving the rural way of life behind.

The family farm was a functioning ecosystem unto its own, but it was not the original ecosystem. Photos from the 1920s show landscapes of fields everywhere—pasture, hayfields, orchard trees, crops, and gardens. Through the span of my father's life, farm after small family farm across the eastern United States was abandoned. A way of life became rarer and rarer. Numerous factors have been invoked to explain this, from the increasing scale of industrial agriculture in the rich soils of the Midwest and California to the efficient transportation network that allowed shipment of produce from afar.

In the same lifetime that measured the decline of family farms, the return of forest cover was dramatic. By the same token, the *potential* habitat for un-

FIGURE 10.4.
Highland Fruit Farm, eastern Ohio, in 2019. Unlike many small farms, much of the original farm remains clear, though only hayfields remain. Steeper portions have been abandoned and show the scraggly forest of early succession. Such forests are poor habitat for ginseng and other understory herbaceous plants.

derstory plants such as ginseng expanded dramatically after 1928. However, the rebound of the understory from agricultural abandonment differs markedly from recovery after timber harvesting. After timber cutting, ginseng and other understory plants could survive the transformation of the overstory because much of the original soil persisted. That soil contained the seeds, roots, rhizomes, and other "propagules" ready to begin regenerating the understory immediately.

Agriculture transformed the soil. Understory plant populations had been reduced to zero. Even the seeds in the soil were gone, replaced by seeds of grasses and weeds. Old fields were colonized by early successional weeds, some native and some not. Gradually, early successional shrubs and trees would seed into the abandoned field, and a forest canopy would form in a process called secondary succession. Early on, these "pioneer" species were most often fast-growing native tree species such as eastern red cedar, Virginia pine, loblolly pine, tulip poplar, and black locust, but in the past several decades they have also included nonnative invasive shrubs (e.g., shrub honeysuckles and autumn olive) or trees (e.g., tree of heaven and princess tree). Vines such as native grapes and poison ivy and nonnative Japanese honeysuckle, kudzu, and Oriental bittersweet took advantage of the opening, clinging to and growing along with the fast-growing trees. After fifty years, such forests often have a scraggly look to them (fig. 10.4). Studies have shown that while tree cover is regained quickly in old fields, recovery of the original tree composition takes much longer. Recovery of the original forest understory is even slower than that. So, while one lifetime saw the recovery of much forest tree cover in ginseng country, it did not span the transition from old field back to the original biodiverse temperate forest ecosystem, with its full complement of plants and animals. Complete recovery, if it occurs, will require many more decades.

Canary in the Coalfields

Underground deep mining for coal would have little effect on ginseng, other understory plants, or the forest cover above, for that matter. But two trends in mining in the twentieth century would change that dynamic. First came strip mining in the 1960s, in which huge machines would clear away the rock layers down to the coal seam, where it could be removed en masse. Even in hilly areas the "overburden" could be taken out, then coal could be removed around the contours of the hill, creating flat benches with high walls that to this day persist across large areas of coal country. Still much coal remained—enough to provide power for centuries to come. The next extraction process that became widespread in the 1990s and continues to this day is mountaintop-removal mining, in which the entire tops of hills in southern West Virginia and eastern Kentucky are removed with explosives and heavy machinery to reach the coal seam below. The mountaintops are then dumped into nearby valleys, leveling the landscape of formerly mountainous terrain. More than five hundred mountaintops have been destroyed by this process, with mines destroying a staggering area of 1.45 million acres of native forest across seventy-four counties in four states.

Obviously, ginseng and other understory plants do not survive surface mining. In many areas, the pulverized rock led to acid mine drainage that inhibited reclamation of the land to forest. Pioneer tree species would often invade and begin to grow, sometimes poorly, on the reclaimed surface mines, but the understory was even slower to recover than after agriculture since the soil layer was starting from scratch.

My PhD student Jessica Turner-Skoff observed that much of this mining was taking place in exactly the same areas of Appalachia where ginseng harvest was the highest. Like me, Jessica was a child of Ohio who had grown up loving the natural world. She impressed me with her fascination with subjects as diverse as pawpaws and kestrels. But what convinced me to take her on as my conservation fellow after Sara Souther graduated was the courage she showed to confront head on the difficult conservation conundrum presented by coal mining in the heart of ginseng country.

Jessica wondered whether ginseng could be restored to forests that had reestablished on old mine spoils. As the centerpiece study of her dissertation, she

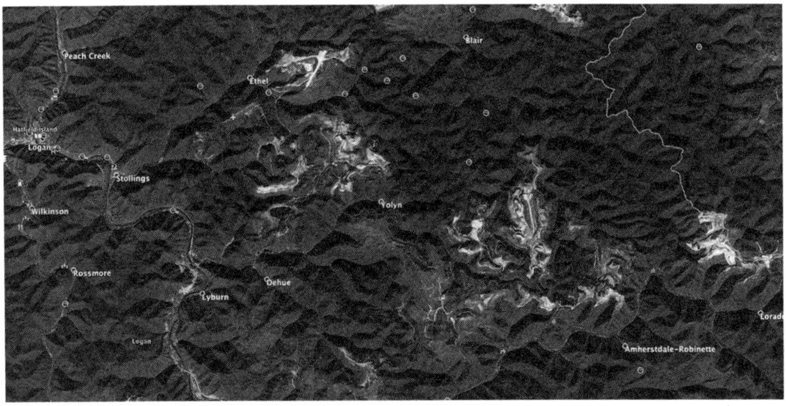

FIGURE 10.5.
Google Earth imagery of mountaintop-removal coal mining operations east of Logan in southern West Virginia

set out a large multiyear experiment to test the idea. In order to gain replication at all levels, her experimental design was quite elaborate. The important point for this discussion was that she incorporated two sites, one in Ohio and one in West Virginia, and each site had three land use history types: (1) mature forests with large trees and a well-developed understory (a kind of control), (2) post-agricultural forests less than sixty years old, and (3) the forested bench of an unreclaimed mine below a highwall that was abandoned less than sixty years ago. In treatments two and three, ginseng had been deleted from the landscape by past land use, but the tree canopy was there. In all, Jessica planted 1,222 ginseng plants, one to three years old, that had been purchased from a wild-derived source in Maryland.

Of course, Jessica measured many parameters to gauge the ginseng's performance over the ensuing three years. All of them told a similar story, so, to illustrate, I am including results for one parameter: relative growth rate (RGR), in this case measured as the change in leaf area of surviving plants (fig. 10.6). If RGR was greater than zero, the transplants actually grew over time.

The results differed dramatically for the two sites. In West Virginia, the soils were naturally more acidic, and both previously disturbed sites (highwall and post-agricultural forest) had very poor ginseng growth. At those places survival of transplants was very low as well. As expected, plants did very well in mature forest at both sites. The contrast between the Ohio and West Virginia sites was for the two disturbed sites. In Ohio, with better soil properties, there appeared to be some hope for reclamation of ginseng plants, but this was not the case on

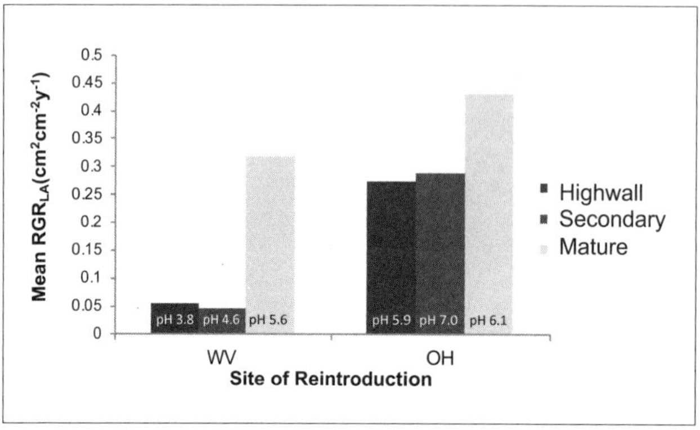

FIGURE 10.6.
Response of relative growth rate (RGR) to site of reintroduction, and land use history (highwall mined, secondary forest post-agriculture, mature forest). On each bar is also shown the average pH. Low pH is associated with acid soils, while high pH is associated with neutral soils having high levels of natural calcium (lime).

the acidic soils at the West Virginia site. This was a positive sign; at least at some older post-mining and post-agricultural sites, restoration of valuable native medicinal plants such as ginseng could be undertaken when the soil allows it.

The Big Picture

If we had Google Earth from 1600 to the present, we could precisely plot the potential habitat for ginseng across the eastern United States versus time. Sadly, our view of Earth from space is a much more recent phenomenon. Some attempts have been made to reconstruct forest cover based on other kinds of data. Figure 10.7 is my attempt to sketch out the rough extent of ginseng habitat using multiple data sources. This should not be viewed as anything other than a very rough reconstruction based on known patterns of land use over time, one that is rougher the farther back in time we go, for obvious reasons.

Given the evidence that ginseng performs best in mature forest communities, figure 10.7 makes clear that a significant part of the decline in ginseng since 1650 can be attributed to the increasing frequency of habitat that is either unoccupied (permanently transformed or agricultural land) or suboptimal (regrowing forest). Thus, in addition to the impacts of harvest, deer browsing, and climate change, we can add habitat loss or degradation to the ever-growing list of reasons for ginseng's decline across the range. As regrowing forest begins to

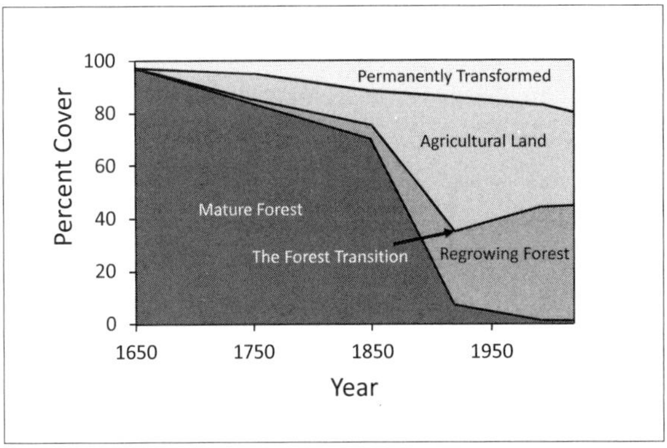

FIGURE 10.7.
Rough reconstruction of percent of the landscape in the eastern deciduous forest ecosystem found in each condition from 1650 to the present. Data drawn from multiple sources.

regain properties of mature forest, with canopy gaps and deep, rich soils, ginseng habitat may increase significantly in the future.

Disturbances to the forest over time provide openings for introduced species to gain a foothold. Some of these become invasive. Can this additional layer of impacts, essentially a by-product of disturbing the landscape, explain some of the additional challenges faced by an understory plant such as ginseng?

CHAPTER 11

Space Invaders (and Other Pests)

We found that the total reported costs of invasions reached a minimum of US$1.288 trillion (2017 US dollars) over the past few decades (1970–2017).—**CHRISTOPHE DIAGNE ET AL.**

Invasives Have It Tough

Hot topics in the field of ecology come and go. In the 1990s and early 2000s, the topic of invasive species occupied a top spot in terrestrial and aquatic ecology. During that period, invasives were studied at many levels, from the molecular up to the entire ecosystem. They were routinely identified as one of the leading causes of extinction of native fauna and flora. In addition, invasive species cost economies billions of dollars annually in prevention and control. As with climate change, overharvesting, overbrowsing, and landscape disturbance, the ultimate cause of the invasive species problem was . . . us.

Species migrate naturally, of course. But there has never been a moment in the history of Earth when a single species has moved around the planet so rapidly, carrying so much cargo. While we have deliberately carried many species from one place to another, recently the primary mode of dispersal has been hitchhiking on all that moving cargo, whether it is in ships, on airplanes, or in trucks and cars. Now species are hurtling around the globe at an unbelievable pace, most of it unbeknown to us until it is too late.

Invasives have a tough go of it at first. Imagine a seed stuck in a small piece of soil on the boot of a tourist who has just visited South Africa and returned by plane to the eastern United States. If that seed falls off in the airport or in the suitcase or the boots are washed down in a sink once at home, the chances of the seed germinating and producing an adult grass or shrub or tree are near zero. The best shot at survival will be if the returning hiker dons the dirty boots and goes for a walk in the park, with the clump of seed-containing soil falling off there. Of course, that is not enough. This has to occur at a time of year suitable for germination in a spot with the right conditions. If it happens to be in a curated space such as a lawn, our little invader will no doubt get mowed down. On the off chance that it is in an uncurated spot, no doubt it will encounter already-established and well-adapted native plants that have a head start. On the now miniscule chance that the seed grows up to produce flowers, it better be self-compatible and not require mates because it is the only member of its species around; otherwise, no offspring will be produced.

With all those low percentages taken together, you might think the chances are so low that a seed from South Africa could establish a plant population in the United States that we really need not worry. But, to take just one year, in 2015 there were 297,000 U.S. citizens that visited that country. Given the high prevalence of seeds in soil, I would guess at least a few thousand of them are bringing back seeds on their shoes or boots. This is happening every year and only needs to achieve success once for an invasion to begin. And remember, I have traced just one pathway from one country to one other country in one year.

Of course, there are many other ways species can disperse beyond their native range. In plants, invaders often hitchhike across the globe by masquerading as useful plants whose properties seduce vulnerable humans into introducing them deliberately far from their native lands for purposes such as landscaping, gardening, or wildlife enhancement. Some of these reside securely within the confines of their introduction sites, but others show no compunction about taking off and proliferating in the surrounding countryside. Not knowing any better, native birds are often accomplices, eating fruits and dispersing seeds far and wide. Indeed, many deciduous forest invasive plants were established this way, skipping right past the low probability of transport by tourists.

Then Invasives Have it Easy

The next problem, which became clear in all that 1990s and early 2000s research, is that invasive plants once they are established can have huge advantages over natives. One hypothesis that appears to explain many invasions is the Enemy Release Hypothesis. By this mechanism, the newcomer establishes itself in a new place, but its enemies—diseases of all kinds, and herbivores—are conveniently thousands of miles away, back home. The invader is free to expand unhindered by, and released from, its usual enemies, and it therefore thrives. Meanwhile, it is surrounded by native species that have their usual complement of pests, thus giving the invader a large advantage in survival and offspring production.

Another hypothesis posits a kind of biochemical warfare carried out by the invader through the production of chemicals in leaves that inhibit growth of competitors. Not having co-evolved with the new plant, our native species are more strongly affected, giving the invader the upper hand.

Yet another hypothesis gives the advantage to the invader because it can exploit resources in new ways, ways that are not taken advantage of by native plants. For example, some invasive species leaf out earlier than the natives and naturally retain their leaves longer, yielding a greater energy gain due to a lengthened growing season. The bottom line is, once established, one or more of these factors may give an invasive plant species the opportunity to grow and proliferate across the landscape rapidly, to the detriment of natives.

Invasives and Ginseng

Kerry Wixted migrated to my lab fresh from an undergraduate degree in conservation from Frostburg State University in Maryland. From early childhood, Kerry waded into streams, turned over rocks to capture slithering salamanders, ran through meadows to catch insects and study them in jars, and picked up turtles and snakes to see what made them tick. I recognized in her someone who had shared my own childhood pursuits. She had the ingredients of a first-rate scientist—natural curiosity and unbounded determination.

In college, Kerry learned that some species "belong here" but others are aliens wreaking havoc on the natural bounty she loved. Taking our ginseng-centric view of the ecosystem, she first posed the most obvious question—"To what extent is America's most famous native medicinal plant exposed to invasive species across its range?" Invading plants tend to do especially well with high resource availability, but low temperatures (e.g., in boreal regions), low water (extreme deserts), and low light (dark forest understories) can inhibit invasions. Given these basic facts, we were not sure invasive species were a problem for ginseng.

The first thing Kerry had to do was to define "exposed to." This could mean an invasive was somewhere in sight and could conceivably spread to be near ginseng within the lifetime of a plant. Or it could mean the invasive's roots are intertwined with ginseng's or its canopy is casting shade over the plant. For practical purposes, she chose to study invasives "actually or potentially interacting with ginseng." The proximity that met this definition depended on the growth form of the invasive. If it was a nonnative tree, Kerry thought a reasonable distance was ten meters (about thirty-three feet). For shrubs and vines, a ginseng plant was considered exposed to an invasive at a distance of five meters. And for herbaceous invasives, the relevant distance was two meters. Of course, these distances are somewhat arbitrary, and results would differ if alternative distances were chosen, but they were considered reasonable "best guesses" about actual or potential competition. Finally, Kerry trained the census takers in our thirty populations to recognize invasive plants they might encounter during their detailed data collection procedures in 2006 and 2007.

Given the relatively undisturbed, low-light environment of the forest understory, we would expect a low invasive species presence there, but we were startled to learn that twenty-one of the thirty populations, 70 percent, contained invasive species within competitive range of at least one ginseng plant. This statistic alone suggests invasive species were a potential problem. Looking more closely plant by plant, fully 37 percent of the 4,279 plants censused in 2007 had an invasive within competitive range. This number ranged between 0 percent and 100 percent, depending on the population. Twelve invasive species were found in total: two herbs (garlic mustard and Japanese stiltgrass), eight shrubs or vines (Japanese honeysuckle, shrub honeysuckle, multiflora rose, Japanese barberry, Japanese wineberry, autumn olive, winged euonymus, and privet), and two trees (Norway maple and tree of heaven). The species number is probably an underestimate since the shrub honeysuckles and privet could have been multiple species. Among these species, shrubs were the dominant invasives, with multiflora rose, Japanese barberry, and shrub honeysuckle ranked first to third in frequency (see fig. 11.1). This may not be surprising since all three are

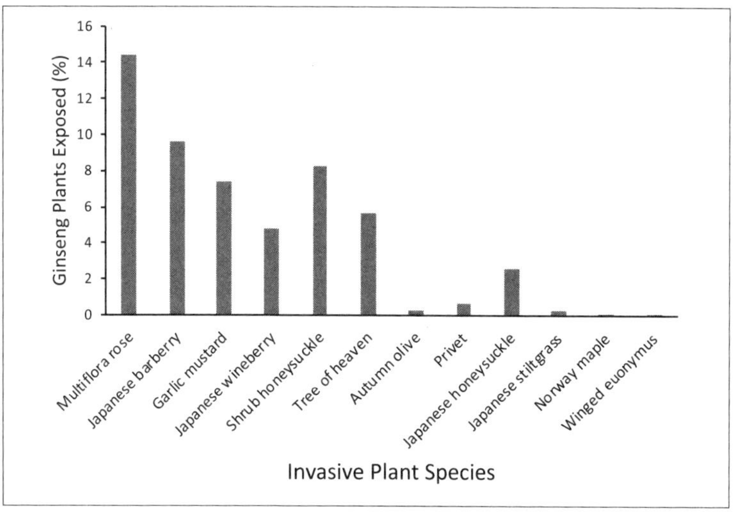

FIGURE 11.1.
Exposure of ginseng plants to twelve common invasive species in the eastern deciduous forest

dispersed by birds, allowing them to invade long distances into the woods from border regions. Among trees, wind-dispersed tree of heaven was close behind the shrubs. Of the two herbs, garlic mustard was the most frequent.

Once the potential for an effect is demonstrated, several approaches are possible to further investigate whether ginseng is competitively suppressed by invasives. One approach would be to remove the invasive or invasives and observe whether ginseng does better in their absence. We didn't want to manipulate our thirty control populations, however. Kerry chose the alternative approach of adding invasive plants to see if they suppressed ginseng performance. Of course, she did this in a population outside the thirty, one we had relegated to experimental status.

Which species should Kerry have added to the mix? For practical reasons, adding trees and shrubs would be quite challenging, so she instead chose to add an invasive herb, garlic mustard. This also seemed like a good candidate because garlic mustard was the most abundant invasive herb near ginseng plants in the forest, and it was known to be allelopathic (i.e., it produced chemicals in the leaves that were known to suppress growth of native species). For comparison, Kerry also added a native competitor to the study, striped violet, which is similar in size to first-year garlic mustard rosettes. Cleverly, Kerry added an activated carbon treatment to the soil in half the competitive setups to absorb possible allelochemicals to see if that reduced the negative effect of added garlic mustard (but perhaps not of the violet?). She hypothesized that the invasive would have a greater suppressive effect than the native but that, if allelopathy was involved, that greater effect might depend on whether activated carbon was in the soil or not. The entire experiment was performed over a two-year period in the forest understory with small replicated competitive arenas using ginseng seedlings.

What did she find? First, competition from garlic mustard and competition from striped violet each increased the mortality of seedlings by about 2.5 times, so in this sense the invasive was not more damaging than the native competitor. However, in a natural population, if the invasive was added to the existing community, this result means ginseng could be negatively affected, at least at the vulnerable seedling stage. The data also suggested that the effect of garlic mustard on mortality could have been due to chemicals it released into the soil, whereas this was not the case with striped violet.

For surviving seedlings, growth variables were almost uniformly better when activated carbon was added to the soil, suggesting that some organic chemical in the soil was suppressing ginseng seedling growth, regardless of

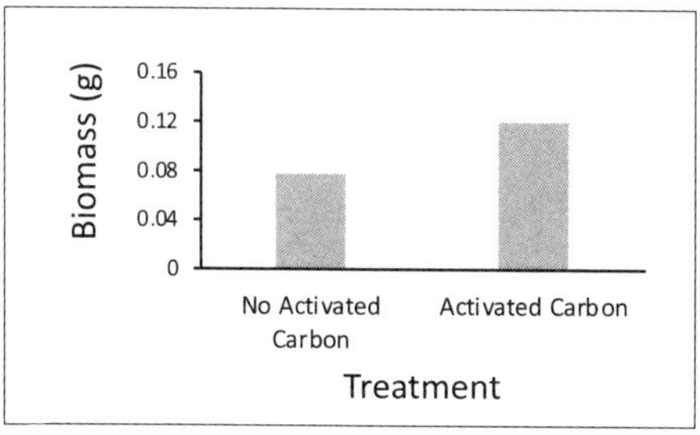

FIGURE 11.2.
Effect of adding activated carbon to native soil on size of second-year ginseng seedings

whether a competitor was there or which competitor was there. For survivor biomass, for example, plants were 50 percent heavier by year two of the study when activated carbon was included in the soil mix (see fig. 11.2). Was there some pollutant in the soil that was suppressing growth of ginseng, regardless of competition? We did not do the necessary detailed soil analyses to find out, but others have shown that activated carbon can have several different effects on native soils, so unfortunately interpreting that treatment is tricky.

Our foray into invasive species research was not over with garlic mustard. To keep funding flowing, most ecologists run their labs on multiple tracks—if one vein of precious scientific gold peters out (or if the topic falls out of fashion), it is helpful to have another one to mine. After all, there were graduate students to support—literally mouths to be fed—by obtaining ongoing funding. My lab had been following the productive vein of invasive species research in parallel with our conservation work. Tree of heaven particularly caught our attention as it was one of the most successful invasive trees in our region of the eastern deciduous forest. PhD student Alyssa Hanna and I were performing studies of tree of heaven population biology to try to understand its success.

Could tree of heaven be affecting ginseng plants in the understory? This was not an idle question since tree of heaven was known to produce a heavy dose of allelopathic chemicals known as ailanthones. In other native plant species, ailanthones had been shown to increase mortality and reduce growth. At the same time, we had noticed that ginseng was sometimes found under native walnut trees, and walnuts are also known to produce strong allelochemicals. This led us to a natural experiment in which we repeated Kerry's study but outplanted ginseng with or without activated carbon under the canopies of walnut and tree of heaven. For good measure, and to add a native non-allelopathic tree, we also outplanted under sugar maples. Alyssa and Kerry hypothesized that tree of heaven canopies would have the most negative effects on ginseng, followed by walnut and sugar maple, but that this pattern would break down

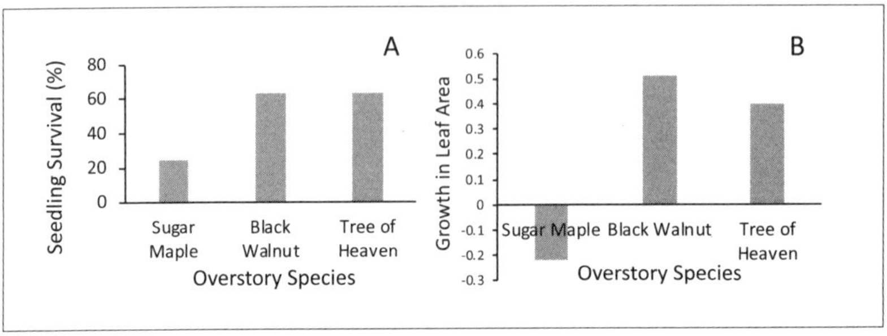

FIGURE 11.3.
(A) Survival of seedlings after two growing seasons; (B) growth in leaf area between 2006 and 2007 under three tree canopy species

in the activated carbon treatments that presumably removed the allelochemicals. Their study included ginseng seeds, seedlings, and adults in order to give a more comprehensive picture of effects.

Ginseng seeds appeared unaffected by the identity of the overstory tree canopy or the presence of allelochemicals. However, we were flummoxed to discover that ginseng seedlings and adults survived and grew well under tree of heaven (fig. 11.3). They also grew well under walnut but relatively poorly under sugar maple. Seedlings were actually shrinking in size under sugar maple. This is odd because sugar maple is often a recommended overstory tree for ginseng growth. What was happening? The short answer: we don't know. Our best guess, which is consistent with other observations, is that ginseng suffers under too much shade and that sugar maple casts the most dense shade of the three tree species chosen for this study. We also speculated that the allelochemicals could have been suppressing ginseng diseases in the soil, which in turn allowed the plants to grow better under the chemical rain from above.

This is where our foray into invasive species effects on ginseng ended. At present, we think invasive species are not a top-three or top-four concern for long-term damage to natural ginseng populations. In truth, however, most of the potential interactions have not been studied experimentally. In some populations with certain invasive species, it seems likely that competition would not be benign. So, much research remains to be done.

Other Pests

In general, we probably understand the least about "pest" species relative to ginseng, including not just nonnative competitors but also diseases. Ginseng growers complain about all kinds of pests—turkeys, voles, and insects, for example. But we rarely saw major problems with any of these in natural populations. One recent compilation of studies showed that there are at least forty-three species of pathogens, mostly fungi, that can cause disease symptoms and mortality in ginseng. We noted a few of these during censuses of our wild populations, but no analyses have been done to examine their overall effects. Only one natural population appeared to show severe disease symptoms resulting in stalk dieback in midsummer in occasional wet years, but roots and rhizomes generally survived. It was probably not a coincidence that this population was the largest and most dense we studied, and it was found on relatively flat ground where soil moisture retention was high. In most of our natural populations, even if symptoms of disease appeared (e.g., small yellow or brown spots on leaves), plants survived and grew normally. Still, did they survive and grow as well as uninfected plants? We don't know.

In cultivation, under shade cloth or tree canopies, fungal diseases are a major hazard for growers trying to bring a given crop to harvest. The cause seems to be primarily the high densities under which plants are grown, although soil amendments such as fertilizer could facilitate fungal growth too. Growers often find they must spray repeatedly and heavily with fungicides in order to save their crop from devastation. In addition, forest growers tell us after they harvest a crop, it is unwise to replant ginseng in the same area. This seems likely due to disease buildup in that area, however, once again, the science has not been done to suss that out.

JGFRT

In our copious field notes taken during censuses, we had to develop shorthand for many symptoms we observed. For example, "YBS" was "yellow brown spot," a leaf condition that probably indicated *Alternaria* fungus infection. Early on, we saw this often enough that it deserved an acronym in the comments section of each line of data. Other observations required years before it dawned on us that we were seeing a pattern. One of these was the gradual disappearance of plants near downed logs or large branches over several years. This contradicted casual observations of the existence of plants emerging from under logs, but the key observation came from changes over several years, showing decline.

What follows is speculation but speculation based on a semireasonable set of assertions. When a tree falls to the forest floor, it takes several years to rot. All kinds of specialized microorganisms colonize the fallen wood and bark to carry out the breakdown process, essentially feasting on the newfound resource. After several years, the log disappears, chewed up by those microbes and released through their respiration as carbon dioxide into the atmosphere. But what happens to all those enhanced microbe populations? As they begin to starve, it seems logical that they would begin to disperse to find new sources of energy. If a ginseng plant is nearby, that woody little root could be just the ticket. Fungi are particularly good at sending out tendrils called "hyphae" in the leaf litter, foraging for more to consume. I spelled out this whole chain of reasoning to a group of students one day after censusing yet another cluster of plants near a decomposing log that was smaller than the previous year. "Jim's Great Fungal Rot Theory," one of them said. "Let's put it in the notes!" And so a new acronym was born: JGFRT. Of course, with hardly a shred of direct evidence for it, we should call it a hypothesis rather than a theory (which has accumulated abundant supporting evidence).

JGFRT. Is it real? Branches and trees fall frequently in the woods, and, no surprise, the persistent ginseng roots will grow a new stalk for a few more years. But the plants shrink and eventually disappear. As for the mechanism, the science has not been done. But just the observation suggests experiments. Would plant groups die out if fungicides were applied to the area? What if logs were experimentally added to some plant groupings but not others, then followed over time? Could the abundance of fungi in the soils, along with the variety, be quantified after treefall? Could fungal disease effects depend on changes in the soil fauna around the log that might damage the root, allowing fungi to colonize? Is the effect the same for all tree species? A dissertation lies in these questions, and the effects could extend well beyond ginseng to other forest herbs. Since trees and branches fall to the forest floor all the time, this mechanism could help explain the spatial distribution of herbaceous plants in the understory. I'm hopeful that JGFRT will one day make it into plant ecology textbooks as more than a hypothesis.

Katabatic Winds

OK, this next hypothesis is really out there, but I mention it here because I really hope someone will test it and show that I am right someday. Because I live on a lake, I have become acutely aware of katabatic winds. During the day, the prevailing atmospheric breezes and winds associated with weather dominate our awareness of air movement. But often these winds die down in the evening. Then a very interesting phenomenon begins. As the sun sets, heat loss from upland areas exceeds heat gain from the sun. The result is that surfaces cool. As the quiet air contacts the cooling surfaces, the air cools as well, gets more dense, and begins to flow downhill in a shallow "wind" concentrated near the surface. This air naturally drains into hollows almost like sheet flow of water over soil. Living on a lake, you can see this wind as small wavelets emanating as a fan out of every drainage. If you walk up the drainage during this time, you will notice the leaves of understory plants waving gently about in response to this force. My far-out hypothesis is that understory plants, including ginseng, do better in micro-sites that experience these katabatic winds because fungal spores are repeatedly blown off their leaves, reducing the chance of infection. These surface breezes may also cool the leaf, which is also not a bad thing on the hot summer evenings when katabatic winds tend to show up.

Other Soil Microbes

We know very little about soils, soil microorganisms, and wild ginseng. Despite measuring basic soil properties in our thirty natural populations, I see little correlation of any one factor with ginseng performance. That's not to say soils are unimportant. Given the general significance of soils for plant nutrition, I have little doubt that they are, but the studies have not been done showing definitively what soil properties are most important in natural populations. It seems reasonable to assume that soil conditions that foster ginseng disease are bad. Likewise, it seems reasonable to assume that organic-rich, moist-but-not-too-moist forest soils are good, while excessively dry or excessively wet soils are bad. There is also observational evidence to suggest that if the leaf litter in the soil is too slow to decompose (e.g., very oaky), ginseng will struggle to germinate and emerge through the thick resulting layer. Others assert that calcium in the soil is a positive indicator, and still others suggest magnesium. There is a lot of speculation and precious little rigorous science behind any of it. How do soils affect relevant parameters such as root shape and ginsenoside concentrations? Are populations adapted to local soil conditions? Ginseng has been shown to have symbiotic relationships with fungi called mycorrhizae; how important are mycorrhizae to plant establishment and growth in the wild? Will climate warming alter the balance of good and bad microbes, tipping the balance in favor of, or against, our beloved plant? So many questions. So much research left to do.

This chapter began with a discussion of "hot topics" in ecological studies. I suspect as molecular methods for microbe identification and quantification become cheap and routine, disease ecology will become a hot topic and attract funding so major progress can be made on questions such as those I have posed above. This is already happening with human biology, as we are only just beginning to understand the significance of the gut microbiome of our own bodies. Why should the microbial environment in and around a ginseng plant be any different? Someday, ginseng doctors may visit a natural or planted population and be able to diagnose health problems like doctors with human patients. Who knows?

CHAPTER 12

Darwinian Forces to the Rescue?

> As many more individuals of each species are born than can possibly survive; and as, consequently, there is a frequently recurring struggle for existence, it follows that any being, if it vary however slightly in any manner profitable to itself, under the complex and sometimes varying conditions of life, will have a better chance of surviving, and thus be naturally selected. From the strong principle of inheritance, any selected variety will tend to propagate its new and modified forms.—**CHARLES DARWIN**

Natural Selection

There was a whole lot of action going on in the world right around 1860. America was on the brink of the Civil War, a conflict that shook the foundations of a nation. And Charles Darwin had just published *On the Origin of Species*, a book that shook the foundations of scientific understanding worldwide. Brilliant scientists stand on taller mountains than the rest of us mere mortals and thus see farther. Darwin understood that the field of geology told the story of an old earth, where lots of time existed for the evolutionary play to produce large effects. He knew from mathematical calculations based in economics that any organism could quickly overpopulate the earth if there were no constraints, but the "struggle for existence" precluded that. He understood through keen biological observations of domestic and wild animals and plants that variation within species was everywhere; you just had to look closely enough. He also knew from animal breeding that there was a resemblance among relatives because traits were heritable, passed down from one generation to the next. These basic knowns of science were assembled by Darwin's Renaissance brain into the radical idea that natural selection, acting on genetically variable populations, was the primary way by which new species came about.

What boggles the mind today, looking back at Darwin, was how much he did *not* know, yet he was able to posit a theory of evolution that has withstood the test of time. For example, though the research was contemporaneous, Darwin was unaware of Gregor Mendel's pioneering studies on inheritance, which demonstrated the mechanism (genes, with two copies at each locus) by which inheritance worked. And only much later would we learn that Mendel's genes comprised stretches of DNA in the chromosomes of cells that were transcribed into proteins that carried out life's most basic processes. And only after *that* would we actually sequence the bits of DNA comprising the genome of an organism and learn how the actively transcribed pieces of DNA are arranged in the chromosome. So the mechanism of inheritance was not known to Darwin; what he knew was that some such mechanism existed because relatives resembled one another. And while geology gave evolution the gift of time, Darwin did not know how much time; only later would precise methods be devised to put an age on rock layers containing fossils and to age the earth itself at four and a half billion years.

Selection is simply the association between phenotype (how an organism looks or functions) and fitness (the relative rate of propagation of an individual's genes into future generations). Differences in adaptive phenotype among populations was what led to ginseng having locally adapted ecotypes in the first place. Do the major stressors we have identified in ginseng act as selective forces? Could ginseng adapt through evolution to the presence of human harvesters? Deer browse? Even climate change? Could this adaptation rescue ginseng from extinction? How could we even quantify natural selection in ginseng?

Selection and Harvest

Emily Mooney asked exactly this last question with respect to harvesters. She took as the target population one of our large ginseng populations that we had been censusing for many years. She then trained four trustworthy pseudoharvesters concerning the laws governing harvest. These people were turned loose in the population for two hours, flagging every individual they could find that they would harvest. The plants were not actually harvested, but Emily could assign a fate of "death" to all the flagged individuals in demographic simulations like those in chapter 5, then observe whether the relationship of phenotype to fitness changed. She was particularly interested in whether there was selection on size of the plant, measured by an index that combined plant height and leaf area.

Emily clearly showed that the relationship between size and fitness could be altered by harvest (see fig. 12.1). Whereas there was strong selection for larger

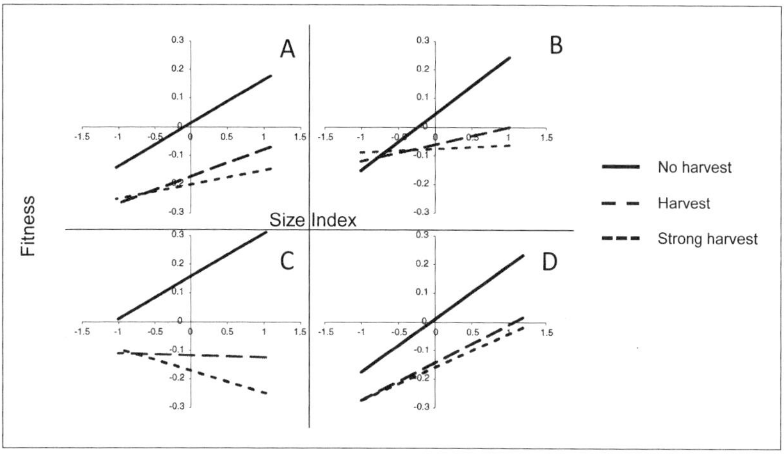

FIGURE 12.1.
Selection pattern for four harvest simulations (A–D) performed in the same population under each of three selection regimes: (1) no harvest, (2) harvest with mortality of harvested plant but seeds left on site, and (3) strong harvest, with both harvested plant and seeds of that plant removed by the harvester

size without harvest, this relationship was weakened or even reversed (simulation C) in the presence of harvest. In practical terms, this result means that over time, if there is a genetic basis for size differences among individuals, selection would no longer favor larger individuals and indeed could result in dwarfing of a population. Although selection by harvesters is only one factor that might be involved, Emily's results could help explain the reduction in size of ginseng herbarium specimens over the past century shown in chapter 2. The result is also consistent with the observation that roots have gotten smaller, and therefore the number of roots per pound has increased (see chapter 2).

Why would wild harvest result in smaller plants over time? The simple explanation is that large plants are more "apparent" to harvesters as they search for ginseng in the woods. Anecdotally, we certainly observed this in our own surveys for ginseng. We would often find a large plant first, and only then notice the many smaller individuals nearby. In addition, harvesters have told us that they will scan the understory for red berries in the fall, and large plants are more likely to bear fruit.

Weakened selection on size, or, even worse, a selection pattern favoring smaller, less apparent plants, could inadvertently produce plants that are less valuable to harvesters. Ecologically, this could affect population viability as well. While harvest pressure could be reduced with lowered apparency (the "adaptive" part of this evolutionary change), the population could be more vulnerable to extinction from other stressors since small plants have higher mortality rates generally, and a smaller seed bank would mean lower ability to recover from a die-off of adults.

Selection by Deer

Deer would be expected to have a very similar selective effect to that of harvesters. While deer do not often kill the plant, at least in the first year of herbivory, they will often eat the fruits and destroy the seeds contained within. Mary Ann Furedi showed that deer preferentially browse large plants, just as harvesters preferentially harvest them, and for some of the same reasons (greater reward per unit effort). Although the effects of harvest and browse are not identical, their effect on fitness is likely very similar.

The Cause of Dwarfing?

Unfortunately, we cannot draw a firm conclusion that the reduction in ginseng size over centuries is due to a genetically based dwarfing process caused by harvest or deer or both. Why not? Because *evolution* by natural selection requires two things: (1) an association between fitness and a trait (we have shown that), and (2) a genetic basis for the difference in size among individuals. The latter has not yet been proven by experiment for ginseng, though there are plenty of examples where differences in size, or differences in growth rate leading to size differences, exist in other plant species. It would not be at all surprising if there is at least some degree of heritability to that trait. But in plants, size responds in a very plastic manner to the environment. So, for example, while deer browse could produce an evolutionary change over generations, the effect of browsing itself will make plants smaller, on average—an "environmental" (nongenetic) effect of the stress of having the aboveground part of the plant removed, as we have seen in chapter 7. Harvest could also have an effect that is not a genetic change over generations; if harvesters simply remove the large plants, the mean size gets smaller, even if there is no transmission of the small-size advantage to the next generation. Perhaps the plants left behind are simply younger and therefore smaller.

Genetic Basis of Plant Traits

We are beginning once again to enter the realm of exciting science left to be done. Good science often leaves more questions than it answers. How heritable is growth rate and size in ginseng? Is it possible that there is a genetic trade-off between the ability to grow fast and production of plant secondary compounds such as ginsenosides? If so, this could help explain biochemical differences between cultivated and wild plants.

We know how to design experiments using controlled crosses to tease these answers out of the data, but no one has taken the time to do so yet. This would require a massive commitment of several years to perform all the crosses, plant the offspring (waiting one and a half or more years for germination), and then grow the plants to adult size.

Sequencing the genome of wild ginseng could help in genetic studies; however, there is no one "size gene" or "growth rate" gene. Instead, traits such as these are known as quantitative traits; that is, they are determined by many genes acting in concert, not just one. The genome of Asian *Panax ginseng* has been sequenced, though as of this writing the sequencing of the sister species, our wild North American ginseng, has not been completed. The genome assembly for both is challenging because they are tetraploid, meaning the whole genome was duplicated at some point in the long-distant past. However, it is just a matter of time before the sequencing of *Panax quinquefolius* is done. Once it is, many more tools for genetic analysis will be available.

The earliest studies of genetic variation in and among populations were done using "isozymes"—protein variants of single genes run out on a gel that reveal what the genetic composition is at that particular spot in the genome. Jennifer Cruse-Sanders cleverly compared protein variation in and among a set of populations, some of which were protected from harvest, some of which were not protected. She found a large amount of the genetic variation was located among populations—that is, particular genotypes were often found in one population but not in another. This "structuring" is what one might expect in populations that are locally adapted, have a breeding system that encourages only local movement of genes, and have some inbreeding. As it turned out, the populations that were protected from harvest had less of this among-population difference, suggesting that exposure to harvest was having effects on the genetics of the wild populations.

After centuries of harvest and decades of overbrowsing, ginseng populations are still harvested and browsed at high rates, so one could argue that natural selection has not rescued ginseng thus far. This does not mean populations haven't changed genetically in response to these pressures—the evidence suggests they have, but the response so far has been inadequate to escape the pressure. As plants have shrunk, both deer and people seem willing to take smaller and smaller plants.

Selection and Climate

What about climate change? Can genetic variation combined with selection rescue ginseng from a rapidly changing climate? Two factors we have already discussed give us reasons to be skeptical. First, ginseng populations appear to be locally adapted to climate at a site. When warm years happen, population growth rates fall below replacement level and populations decline. The second factor is the relatively low within-population genetic variation we just discussed. Unless there is a subset of individuals within populations that are genetically programmed to survive, grow, and reproduce better under warmer conditions, natural selection cannot easily come to the rescue of ginseng.

To test the idea that relevant variation might exist, I pulled out data from the natural population of ginseng we had been censusing the longest (Population 27). Knowing the mean maximum daily temperature from each growing season, I then asked whether individual plants differed in their response to the temperature variation from year to year. To do this, I found all the individuals that we had followed for at least seven contiguous years *and* that had not been browsed during that time. I excluded the browsed individuals because we knew this would simply add noise to the data—browsing would reduce growth regardless of temperature—and we were looking for what might be a very subtle signal. It turned out that 252 individuals had multiyear records that matched the criteria. The statistical test showed just a trend suggesting that individual plants might be responding differently to temperature variations. That sample size gave quite a lot of statistical power to find differences if they existed, yet there was no significant difference in the strict sense. This did not lend much hope to the idea that populations contain important variation in ability to respond to warming temperatures.

IMHO

Having studied ecological genetic variation since the beginning of my career, I will be one of the first to defend evolutionary processes as extremely powerful forces for change, even over short distances and short time spans. In this case, however, the time frame may be just a little too short, relative to the life span of individual plants. And because of ginseng's breeding system, combined with the isolation of populations containing genes different from one another, the cards are stacked against genetic rescue by natural selection.

Could there be another mechanism by which ginseng could heal itself ecologically, in the face of all the insults we humans have hurled at it? We thought we might have found an answer in the call of my favorite spring bird song—*EEE-O-LAY!* The next chapter unravels the mystery.

CHAPTER 13

Dispersal to the Rescue?

> This is the only bird whose note affects me like music. It lifts and exhilarates me. It is inspiring. It changes all hours to an eternal morning.—**HENRY DAVID THOREAU**

Why Disperse?

Dispersal is nature's insurance policy against a degrading environment. At a grand scale, dispersal expands a species' range. At a local scale, a plant degrades its own environment by locally depleting soil resources that it requires to sustain growth and reproduction. Thus, baby plants, in the form of embryos within seeds, need to escape the zone depleted by their mothers.

Beyond resource depletion, maternal plants can, over their lifetimes, attract a host of disease or consumer organisms that they might withstand, at least for a time, but would be lethal to their vulnerable offspring growing nearby. This mechanism has been invoked as a major cause of the high diversity of tropical rainforest vegetation; individuals of the same species just don't readily establish next to their maternal parent. Dispersal also helps ensure that when the baby plant grows up, it will be less likely to mate with relatives (inbreeding), and have more fit offspring as a result.

For a ginseng plant, a human brain's memory of a plant's location can mean a degraded environment as well, in the form of a greater risk of death by harvest. Dispersal to a location not found in some human's memory could give a young plant much better prospects for long-term survival.

Dispersal could also rescue ginseng from deer. Deer follow well-worn paths through the forest, and they will often browse near those paths as they traverse the landscape. Dispersal could provide escape from proximity to one of those paths, even if only for a few years. In addition, certain micro-sites are avoided by deer, and dispersal to those areas can mean a safe site for young seedlings. With severe deer overpopulation, escape from browse may be nearly impossible, save for those rare sites with perched rocks that cannot be reached by our four-legged friends (see chapter 7).

Our other top-three stressor, climate change, may put the largest premium on long-distance dispersal. Climate change is moving the optimal climate for locally adapted populations up in elevation and north in latitude. Since individual plants are rooted in place, the best hope for following the optimal climate may be something that could move genes upward and northward.

Dispersal in a plant like ginseng can occur in one of two ways. First, pollen can disperse from the anthers of a flower on one plant to the stigma of a flower on another plant some distance away. But how far away? This was one of the first questions we asked in my lab when we first wanted to know how to define a population of ginseng. The second means of dispersal, dispersal by seeds, was the subject of one of the last studies we did.

The Wagon Wheel Experiment

Master's student Erin Hackney and I designed an experiment to see just how far pollen can flow, using actual ginseng anthers and stigmas to measure the rates. We called it "the wagon wheel experiment," and at the time I thought it was brilliant. In three separate ginseng-less forests, Erin transplanted two sets of adults in clusters at the center of the wheel (see fig. 13.1). One central cluster was small, and one was large. Then she planted additional transplants at 16-foot intervals along spokes in eight cardinal directions away from the center population, all the way out to 160 feet away. Finally, for all plants along the spokes, as flowers matured, she clipped off the anthers—the pollen-producing organs of the flower—to prevent self-pollination. This ensured that pollen would have to be transported there from the central cluster of plants to successfully sire viable offspring. After flowering was complete, she allowed berry formation to proceed and observed where along the spokes berry production began to decline. Those distances would mark the "end" of the population with respect to pollen movement. The design of this experiment may seem simple, but it was a lot of work since Erin had to visit every plant several times a week for over a month in order to catch flowers at the right stage.

The results were never published, but I describe them here to illustrate what we did learn and to show how seemingly elegant experiments often fail, or at least give ambiguous results. Upon reflection, this particular experiment may not have been so brilliant.

First, we discovered that not all transplants are successful; only 60 percent of the transplants emerged when temperatures warmed in the spring, substantially reducing our sample size. Also, when this experiment was done in 1999, we had only just begun to see how impactful deer were on ginseng in

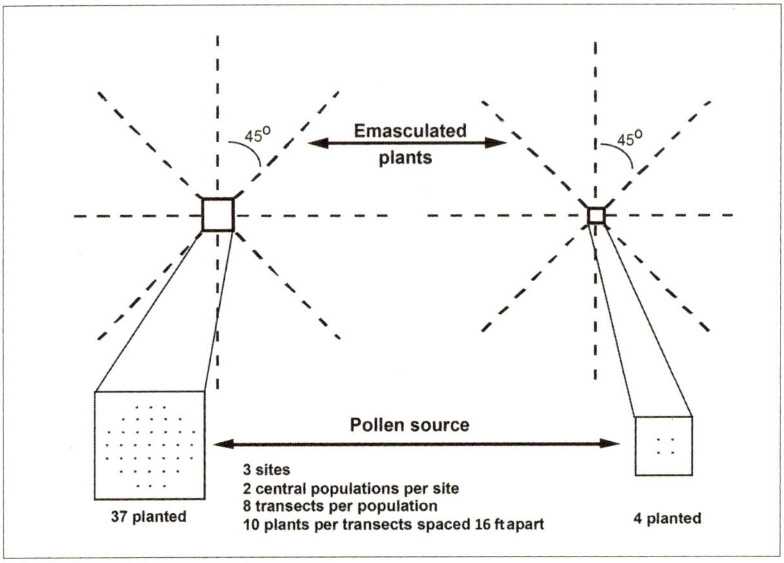

FIGURE 13.1.
Schematic of the "wagon wheel" experiment designed to measure gene flow mediated by pollen from a central pollen source population. These arrays were repeated at three field sites.

the woods. For plants that emerged, about 40 percent of them were browsed and therefore not useful in the experiment because they had no flowers. Both of these factors introduced missing values to certain distances from the center and altered the effective size of the central pollen donor population. Nevertheless, we attempted to glean what we could from the remaining plants.

At two of the three sites, there was no falloff in seed production with distance, while at one site, fruits per plant and per flower did decline with distance. This suggested that pollinators may have been limited in their ability to carry pollen at one population but not the other two. But in this experiment, every ginseng plant was potentially within sixteen feet of another ginseng plant, so perhaps it was not really testing genetic isolation at all. Some pollinators are known for a behavior called "traplining," where they essentially follow a trail from one individual to the next. Our spokes could have been de facto trails for pollinators. For pollen to reach the outer end of each spoke, it is true that pollen would have to be carried 160 feet, but would a pollinator have gone that far if there were no ginseng plants in between? We don't know. Given the resources, time, and effort that went into this experiment, Erin's committee did not withhold her degree because the experiment was "imperfect" (thank goodness), but direct observations of pollen movement were not attempted again.

The Allee Effect

Erin did perform a more conclusive experiment examining the effects of ginseng cluster size on seed production and ultimately population viability. The results of this study were indeed published, and because of its implications it has been referenced by other scientists over one hundred times.

Once again using transplants from cultivation, Erin placed 596 plants out in the woods in twenty-six isolated groupings ranging in size from one (N=1) to fifty-two (N=52). She subsequently made pollinator observations on the groups of different size and again observed fruit and seed production. She reasoned that pollen donors (males) would be less limiting in large groups than in small groups. Therefore, if pollen flow was important to overall seed production, she would expect a falloff in seed production at low N and increase in seed production at high N.

Typically, when densities increase, we expect processes such as competition and predation to increase, negatively affecting the fitness of individuals. But Erin was proposing a positive effect of higher density, or what is known as an Allee effect. This had been shown with plants before, but we were not at all sure, given ginseng's ability to self-pollinate, that more outcrossing would improve fitness.

Indeed, ginseng does appear to do better, seedwise, when found in larger clusters (fig. 13.2). Across the range of population sizes in Erin's experiment,

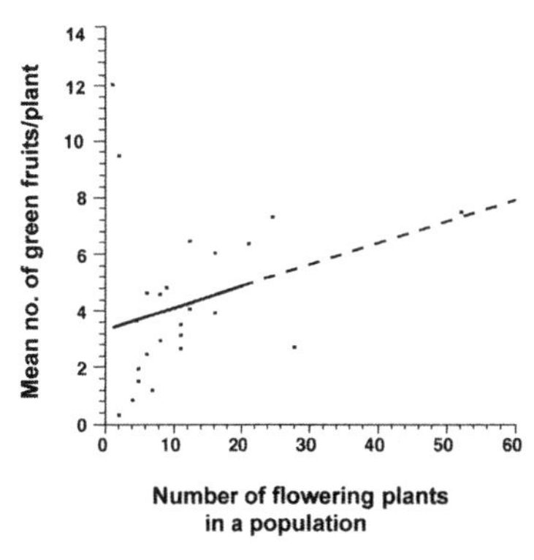

FIGURE 13.2. Increased seed production with greater population size, suggesting an Allee effect due to pollinator limitation at low population sizes

this meant a doubling of the average fruit production from small to large population numbers. When we plugged this level of variation into demographic models for populations, it suggested that, in at least some populations, this level of Allee effect would be enough to drop a population below replacement rate when numbers were small. Erin's pollinator observations showed a large amount of variability in visitation rates, but plants that were visited more often generally produced more seeds.

Ginseng Stripes

Ginseng clusters are not randomly shaped. Typically, they occur in stripes, oriented up and down the fall line of a slope, especially the steeper the slope. Ecologists are fascinated by patterns such as these. If we can understand them, it brings a little order to the chaotic jumble of organisms in nature. In our fall censuses, we often observed that ginseng's bright red berries ended up right at the base of the mom plant, little red beacons in the otherwise drab leaf litter. They simply fell off. Sure, seedlings occasionally germinated (21–45 months later) right there where they fell, but often germination occurred some distance away. How did those seedlings get there? If the slope was steep enough, we imagined that anything that ruffled the leaf litter—wind or animal—could end up bringing the seed downhill by gravity. Hard rains might do the same thing. But how far could they go? This so-called secondary dispersal would seem to have limited power. We wouldn't know without doing an experiment.

Throughout my career, I've been accused of goading my students into doing massive experiments with huge sample sizes beyond what is reasonable. But sometimes students come up with these monstrosities on their own. Martha Van der Voort, in addition to studying harvest effects, was deeply curious about many basic ecological questions about ginseng. It was hard to contain her enthusiasm. So, when we discussed trying to understand gravity dispersal, she designed an experiment—using sixteen thousand berries. Without going into all the experimental details, the crux of the design was that she laid out red berries across the slope at regular intervals between two stakes. For the next four years, she returned and noted all the seedlings that emerged, including their positions relative to the original line she had set out.

The results were illuminating. Fully 90 percent of seeds that germinated did so within two meters (six feet, seven inches) of where they were sown (fig. 13.3). But other processes transported seeds downhill, occasionally more than one hundred feet. We isolated two strong suspects for this medium-distance, directional dispersal. One was water. We noticed that some of the seeds appeared

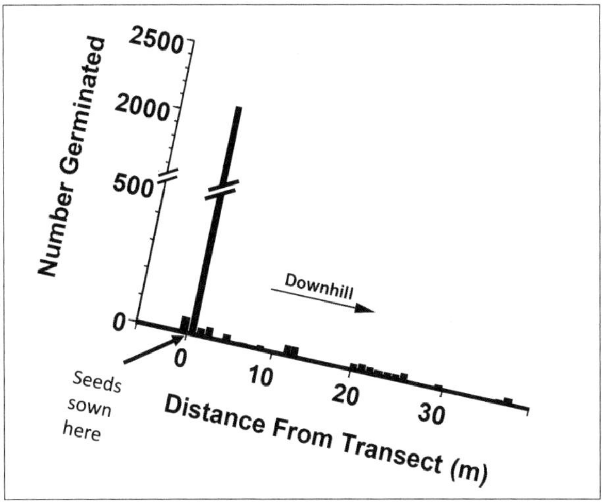

FIGURE 13.3.
Postdispersal movement by ginseng seeds

to be in washes, slightly lower areas in the sloped landscape where water could flow across the surface in heavy rains. This is still gravity dispersal, since gravity pulls water, which in turn pushes or floats seeds. The second suspect was animals. In a couple of cases, groups of seedlings would pop up tightly clustered together, suggesting an animal had cached them there, then forgotten about them. What animal? Another mystery.

Martha's work suggested that the vertical "striping" of clusters on hillsides was indeed largely a product of gravity dispersal at work. Thinking about the spacing of clusters on a hillside, the distribution of wild ginseng made more sense now in light of her findings.

A piece was still missing from the puzzle, however. Gravity moved seeds downhill. Over decades or centuries, one could imagine a general hopscotching of a population down toward stream bottoms. Indeed, one of our thirty populations (Population 20) was precariously perched on the edge of a stream at the bottom of a steep slope. In wet years, while censusing this population, careless undergraduates would accidentally disperse themselves over the bank into the wetness below. It seemed highly likely that many if not most of the ginseng seeds produced in Population 20 suffered the same fate and were lost to the aquatic abyss. Was this population doomed? How could seeds be dispersed uphill? Thinking more broadly, how could new populations get established some distance away from an existing one? Gravity could not be the answer.

In Search of a Mutualist

Research, animated by exciting questions, can start out fun, but it has been known to lose its glamour just a tad after hundreds of hours of grueling fieldwork under conditions ranging from cold and wet to scorching and dry. Certain studies are just plain fun from start to finish, though. And that was definitely the case when my lab started a little side project to discover the identity of ginseng's unknown nongravity dispersers.

In another research life, I got involved with remote sensing—the use of image-collecting devices deployed in drones or aircraft or satellites to observe the forest below. Remote sensing is data collection at a distance. If animals were involved in ginseng seed dispersal, it seemed a reasonable assumption that we would need to separate the human observer from the fruit-producing plants in order to assure normal animal behavior. In other words, we needed to employ remote sensing. Luckily, when this approach occurred to us, a technology was being developed to capture images of wildlife, so-called trail cameras. We had already deployed them to discover what was munching ginseng plants (see Mary Ann Furedi's work, chapter 7). Now we hoped to use trail cams to discover who was gathering the berries, whether directly from the plant or from the ground underneath. Sara Souther and I first took up this cause in the mid-2000s. Based on the diversity of animals that showed up in our frames, it was clear this was more than a side project. We needed a dedicated student to really focus on this problem and attack it more systematically. Amy Hruska showed up in my lab, fresh from an undergraduate degree at University of Dayton, in the nick of time.

Amy listened to my sales pitch about taking on the ginseng-disperser quest and quickly saw that it fit her to a T. The multitalented Amy could wisely see that the study would have to involve much more than gathering imagery of animals nosing around ginseng plants. That would only tell you so much, but that's where the project had to start. To that end, we multiplied our camera inventory and set up camera traps by affixing them to camouflaged two-by-four stakes carefully positioned three to five feet away from fruiting ginseng plants (see fig. 13.4). The cameras were programmed to grab three images when something moved within the frame. In the end, Amy examined 2,775 images snapped by the cameras with 241 "camera days" of data collected from fifteen cameras.

More than ten species of mammals found their way into the camera frame to have their pictures taken. There were multiple species of mice that could not readily be identified. The most common mammal by far was the eastern chipmunk. Fortunately, in combing through the thousands of images captured by

FIGURE 13.4. Trail camera setup for capturing images of ginseng berry munchers

our cameras, Amy had concocted a scheme with five levels (from low to high) for judging the level of interaction with the ginseng berries in order to discern most likely candidates for further investigation. Among mammals, only the chipmunk was found to have high levels of interaction, suggesting they may have been removing and dispersing berries (fig. 13.5A). Chipmunks practice "scatterhoarding," whereby they cache abundant fruits, which may be carelessly forgotten; much later, tight groups of seedlings emerge together. Amy had found the most likely cause of "caching" seen in Martha's study, and occasionally observed in wild populations as well.

This does not preclude the possibility of other species occasionally eating ginseng fruits and dispersing seeds. While Mary Ann Furedi's feeding experiments had excluded deer as possible dispersers, we strongly suspected omnivores such as raccoons and black bears might occasionally disperse seeds. Based on knowledge of their diets, eating bright red fleshy fruits would be likely, and given how often the scat of both species is loaded with seeds, passage through the gut also seems likely. But in Amy's studies these were not frequent events.

FIGURE 13.5.
Animals showing high-level interaction with ginseng berries in trail camera photos: (A) chipmunk, (B) wood thrush, (C) hermit thrush, (D) Swainson's thrush

Bears were unwilling participants in this study; one of them detected our setup and decided to remove the camera from the stake and smash it to bits. The camera recorded all of this prior to its demise. Nevertheless, in future work, it might be worthwhile to pursue the bear dispersal angle, since they may wander the woods for miles in search of food and carry their eaten seeds just as far before defecating them in a fertile spot on the forest floor. In addition, bears are becoming more abundant in the eastern deciduous forest in recent decades. If they are ginseng seed dispersers, they could be important ones.

We still seemed to be missing a disperser capable of moving seeds over long distances. But those who study seed dispersal for a living have long noted that birds perform this service quite well and in fact prefer fruits that are red or black. It is no coincidence that raspberries and blackberries are these colors. These plants have conscripted birds to perform dispersal for them. Likewise, spicebush, dogwood, baneberry, jack-in-the-pulpit, black cherry, elderberry, and a host of other species sport berries in one of these two colors. Do birds also disperse the plump red ginseng berries?

Amy found nine bird species that wandered into the photo frame of our cameras. But using her interaction criteria, just three showed frequent high-level interactions, and they were all relatives: wood thrush, hermit thrush, and Swainson's thrush (fig. 13.5B, C, and D). We strongly suspected gray-cheeked thrushes and veeries would also have affinities for ginseng fruit clusters, but we never caught them on camera.

I have to confess an unscientific bias: thrushes are some of my favorite birds in the universe. I'm with Henry David Thoreau, who was exhilarated by the song of the wood thrush. Up close the wood thrush is beautiful, but it is its song that is simply stunning. If you want a mind-blowing experience, listen to the wood thrush song slowed down on YouTube.

Friend or Foe?

So we had to know. Were those musical thrushes ginseng's friend or foe? Yes, our cameras showed they ate the fruits. But if the seeds were crushed or if they died in the gut as they did inside white-tailed deer, then thrushes were predators. On the other hand, if seeds somehow survived this interaction, then thrushes would endear themselves even more to the botanist in me.

Do you know how hard it is to find captive thrushes to feed ginseng fruits to? Very hard. It is not easy for anyone to keep wild birds, both for legal reasons and animal care requirements. But Amy was determined, and at long last she found a source at the Tennessee Aquarium in Chattanooga. After securing permission from the generous caretakers there, she assembled a team to place ginseng berries in the "Delta Country Exhibit," a natural habitat simulation indoors containing several native bird species. Fortuitously, the aquarium also had caged individuals of hermit thrush, Swainson's thrush, and veery. Wood thrushes were studied in the exhibit along with other birds.

"YES!" was my response when Amy modestly but proudly reported back with her results. Every thrush species ate ginseng fruits, and all of them exhibited the same behavior. The thrushes gulped down the berries whole. Then, somewhere between fifteen minutes and thirty-seven minutes later, their heads and neck would convulse a bit and *boink*, out would pop a fully intact, regurgitated ginseng seed, stripped of pulp. Amy gathered up these seeds and later tested their viability in the lab. They were 100 percent intact and viable. We had found our mutualists.

You already know, of course, that our research would not stop with that oh-so-satisfying answer because our original questions had to do with ginseng dispersal. How far from the mother plant could a thrush take a ginseng seed in the fifteen to thirty-seven minutes when the seed resided inside it? Could it help ginseng disperse to a place not in a harvester's memory, away from deer trails, up mountains, or significantly northward? The best way to find out would be to watch actual thrushes move about the woods at the time of year when they would be gulping down ginseng fruits.

I had to look no farther than my pool of undergraduate conservation interns to find a student willing to take on that project. Michael Elza grew up in the mountain town of Elkins, West Virginia, hunting, fishing, and generally spending as much time as he could outdoors in the summer. As a bright undergraduate science major at WVU, he took the path less traveled—double-majoring in wildlife and biology and taking ecology courses. He had been helping us census ginseng for two summers just because he wanted to be outdoors, and he was one of our most reliable field hands. But he was also a bird-watcher. So when the opportunity came to track wood thrush movements, he jumped at the chance, and he eventually wrote a superb honors thesis on the topic. Mike enlisted the help of Wildlife and Fisheries Resources faculty and grad students who helped him mist-net two wood thrushes and then place tiny radio tags on them. The tags helped him locate the thrushes for tracking when he returned at later dates. Once he located them, Mike would carefully record their movements for set periods of time. Using each point along the track as a possible ginseng location, he could then measure the distance from that point to future points as a function of time to see how far, in theory, an ingested seed would travel while inside the wood thrush gut.

Mike found that most dispersal distances were less than 150 feet, although occasionally a thrush went as far as 300 feet within the time frame before seed regurgitation we had observed previously. The modal (typical) distance was 30 to 60 feet. The fascinating thing about that distance is that it appears to correspond closely to the distance among clusters of ginseng plants within populations. Could the spatial dispersion of clusters be dictated by the thrush digestive system? It seems like a possibility, though it would be great to do some spatial modeling, knowing what we know now, to test that idea further.

Though Mike's short study did not show potential seed dispersal distances of more than three hundred feet, investigation of thrush movement patterns by other ornithologists suggests that greater distances are possible. Juveniles in particular are known to fly a few miles at that time of year, dispersing from their native range. Perhaps it would be rare that munching of ginseng berries would occur simultaneously with long-distance flight, but, as with long-distance dispersal by black bears and other mammals, these rare occasions could be important in transporting genes at a higher rate across the landscape.

In the fall, as ginseng berries are ripening, thrushes are gearing up to migrate south for the winter. Some end up along the southern coasts of the United States, while others cross the Gulf of Mexico and end up as far south as Central America. Unfortunately for ginseng, if a seed happened to end up inside a thrush beginning its southward journey, it would be going the opposite way needed to pursue the optimal climate.

In summary, thrushes and ginseng are mutualists. We have been focused on benefits to ginseng so far, but what about benefits to the thrushes? They gain a food source in the fleshy fruits they swallow. But could they be gaining something else? I ask the next question knowing full well how speculative it is. By consuming ginseng berries, could thrushes be gaining the energy and stamina for long-distance migration, much as humans have used ginseng for thousands of years as an adaptogen to give strength and vitality? Could ginsenosides be their superpower? Could this explain the thrush's attraction to ginseng berries, just at the time of year when they need stamina for a thousand-mile journey? These questions are unanswered, but I would not be surprised to learn that this mutualism is even more fascinating and important than we currently know.

The bird-ginseng mutualism may have been even more central to ginseng's welfare in past centuries. My colleague Daniel Gagnon speculates that the now-extinct passenger pigeon may have played a key role in long-distance dispersal of ginseng, along with many other species it was known to consume. What's more, passenger pigeons were once the most abundant bird in North America, with a conservative population estimate being three billion to five billion in the 1700s. Due to overhunting at their group nesting sites, they were driven to near-extinction by the late 1800s, and the very last bird died in the Cincinnati Zoo in 1914. When they were a major player in the ecosystem, they consumed many species of nuts but also fleshy fruits such as berries, including, for example, dwarf dogwood, whose berries closely resemble those of ginseng. It is therefore almost certain that ginseng fruits would have been part of the passenger pigeon meal plan. Perhaps ginsenosides in the berries fueled passenger pigeon migration as well. Events such as a pigeon swallowing a fruit, then flying a great distance, then eliminating the viable seed from the crop onto the ground, as thrushes do today, were undoubtedly rare. But multiply any rare event by billions of opportunities, and they suddenly become quite common. It is possible that ginseng is missing its most prolific and important disperser entirely, due to extinction.

Three Billion Canaries?

In a massive review published in *Science* in 2019, Kenneth Rosenberg and colleagues estimated the loss of total bird numbers across North America at nearly three billion over a period of forty-seven years. This is a 29 percent net loss

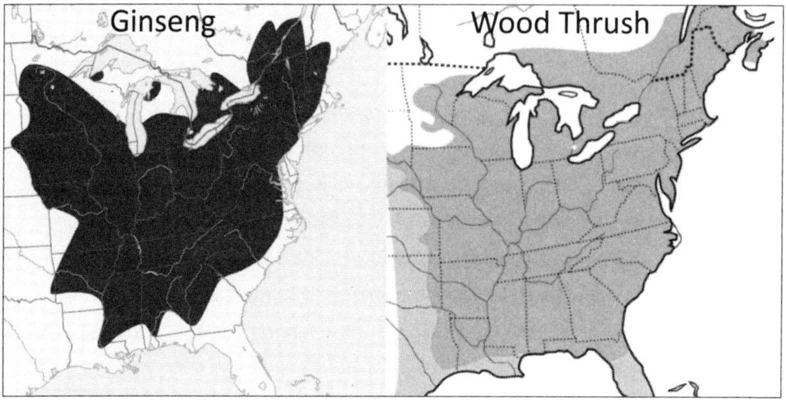

FIGURE 13.6.
Range maps of American ginseng and wood thrush (summer breeding). Images from Wikipedia, Cornell Lab of Ornithology

FIGURE 13.7.
Mutualists. Watercolor by Susan Bull Riley.

from 1970 to 2017. Drilling down through those staggering numbers, I was curious about ginseng's mutualists, the thrushes. Three of the four species we investigated have declined, with the average for all four being a loss of 20 percent. Most worrisome: wood thrushes have declined 55 percent in less than half a century, from 27.4 million birds down to 12.2 million. Of the thrushes, the wood thrush breeding range overlaps ginseng's range almost perfectly (see fig. 13.6), and the wood thrush is likely to be the most important disperser today.

What happens when a plant loses its primary mid- to long-distance disperser? What has already changed about ginseng dispersal with fifteen million fewer wood thrushes now than forty-seven years ago? Will ginseng patches become denser (and perhaps more prone to disease or more prone to being found by harvesters or deer)? Lacking dispersers, will the rate of new cluster and new population formation decline? Will ginseng become less able to track the shifting climate? What happens when the disperser loses its primary source of energy just prior to a marathon migration? Are the fates of these two Appalachian icons, the wood thrush and ginseng, intertwined?

Some may think I am an odd bird. So be it. But I have to admit, every April I hold deep within me a palpable tension until I hear the first wood thrush song in our forested backyard. Could this be the first year in a future of woods without the wood thrush? A silent spring? I cannot bear the thought. My eyes get misty at the notion that my great-grandchildren might never apprehend what Thoreau and I waited for each spring. Then I hear it. First the call. *Bup bup bup*. Could it be? Then the distinctive song, *EEE-O-LAY*. Yes. All is well with the natural world. Ginseng's friend has arrived home. My anxiety has been relieved. Until next spring.

CHAPTER 14

Conservation in the Anthropocene

> The Holocene has ended. The Garden of Eden is no more. We have changed the world so much that scientists say we are in a new geological age: The Anthropocene, the age of humans.—**DAVID ATTENBOROUGH**

Plant with a Brain

If you were a ginseng plant with a brain, you would live in mortal fear of humans. As an embryo formed on your mother's flowering stalk, you might worry that no wood thrushes would pick you up and carry you away from your mother's fungus-infected root. After all, humans have destroyed wood thrush habitat both in the eastern United States and in their wintering grounds in Central America, so their numbers are way down. As a seedling, you might emerge too early in the spring and find yourself rocked by frost damage. Human-caused emissions of CO_2 have produced the expected greenhouse effect, warming winters where you live, giving a false signal that it is time to grow. As a two-leaf juvenile, you would fear that a human harvester would dig you up, even though you were below legal age for harvest, because humans do that, knowing they won't get caught. As an adult, you would tremble at the sight of so many deer wandering through the woods because humans had removed so many natural deer predators and hunting is managed for high deer populations, not ecological balance. And if you managed to live a decade or two, you would worry that your protective overstory tree canopy would be clear-cut or a housing development or strip mall or strip mine would destroy your protective forest cover.

For good reasons, scientists are taught not to anthropomorphize—not to turn our subjects into little humans that perceive the world like we do. This can lead to sloppy reasoning about what constitutes success and failure in the natural world. So, yes, we know that the "ginseng plant with a brain" is absurd and that a real-world plant fears nothing.

But taking a ginseng-centric worldview is valuable for conservation (see fig. 1.1). It promotes empathy for one other species with whom we share this planet and awareness of the interconnectedness of humans to all species around us, even a little ginseng plant. The ginseng-centric worldview flips the script of our usual human-centric world. It gets us over ourselves, if only briefly. At the same time, it makes us aware of ourselves, of our unwitting power to impact other species. Ginseng plants don't actually fear us. But, in the Anthropocene, they should.

Lingchi

Ginseng is not the polar bear. Polar bears rely on drifting sea ice as a base for catching their favored prey—seals. Climate change is melting the ice. Lacking ice, lacking access to seals, bears starve. Simple story. While the polar bear story is more complex than that, its essence remains true. What we are doing to the climate may well result in the extinction of a very beautiful, large, warm, fuzzy white bear.

Ginseng is not the vaquita. Vaquitas are unique, small, cute porpoises that live only in the Gulf of California. Fishermen place nets in their home territories. Vaquitas get caught in the nets. Humans harvest the seas. Vaquitas die. The species is almost extinct. This story actually gets a bit closer to the ginseng story. Fishermen in the Gulf of California are specifically seeking the totoaba, a fish whose swim bladder is highly sought for the Chinese dinner table and is worth $10,000.

Ginseng is not the northern white rhinoceros. Rich Asian buyers who have faith in traditional medicine believe rhinoceros horn has special curative powers. Poachers in Africa shoot rhinos for their horn, leaving the remaining carcass to rot, then sell the horn on the black market, again for astronomical sums. The northern white rhinoceros is virtually extinct.

These conservation stories are neat packages wrapped up with a bow. Easy to understand. Easy to digest. Easy to be enraged by. Easy for nonprofits to use to stimulate donations to fight for their cause.

The ginseng conservation story is not so simple. Ginseng is not warm, fuzzy, or cute. It is not even recognizable to most people. And instead of one human impact, we have layers upon layers of impacts, many of them subtle and requiring years of data to reveal.

With ginseng, we have death by a thousand cuts: lingchi. Lingchi was a form of torture used in Asia as punishment for especially heinous crimes, such as treason. Without dwelling on its gruesome details, suffice it to say that any one cut was not enough to kill, but over time enough cuts would do the job. Death was slow and agonizing. Ginseng's demise is also slow and agonizing.

Lingchi was intentional punishment. The tragic part about our lingchi relationship to most wild species is that it is seldom our intention to inflict those cuts. Instead, it is a by-product of our staggering numbers (approaching eight billion) multiplied by an increasing resource consumption per individual, compounded by our tolerance or ignorance of the unsustainable methods of obtaining and disposing of those resources.

Ginseng has given us the world's prime example of unintended consequences resulting in "death by a thousand cuts." In all the public meetings about ginseng, in all the talks I have given, in all the conferences I have attended, in all the emails I have received about this inconspicuous little understory herb, I have never gotten the slightest inkling that someone would like to see ginseng go extinct. Indeed, in terms of conservation, that is one of the greatest assets ginseng has going for it—we all want ginseng to survive and thrive. Scientists, environmentalists, land managers, landowners, consumers, and even harvesters want ginseng to be around a few centuries from now. Yet we have clear evidence for rapid decline in natural populations throughout the range.

Let's briefly review humans' unwitting assault on ginseng. First, there was harvest. For nearly three centuries, unregulated harvest spread across the landscape, yielding the first great population crash. Simultaneously began the gradual but inexorable loss of habitat through logging and land clearing for farms, resulting in the second source of great decline. In recent decades, deer browsing exploded as wildlife management catered to hunters' desires for abundant prey. Dozens of invasive species are spreading into forests, and, while not all of them are a threat, probably some of them threaten specific ginseng populations. Climate change in the form of hotter summers with more variable precipitation, including stronger storms but longer droughts, is having an effect already and is predicted to worsen. A lot. Coming full circle, despite imposition of regulations on harvest, illegal and damaging harvest is widespread.

Then there are numerous other factors that we suspect are a negative, but the science remains to be done. For example, acid precipitation has lowered the pH of soils across ginseng's range, simultaneously stripping soils of positively charged soil nutrients. This likely slows growth rate and may transform certain habitats from marginal to uninhabitable. Nonnative earthworms churn forest soils and make them less suitable for herbaceous plant growth. Ozone and other air pollutants exact a biochemical toll on leaf structure and physiology. Nonnative fungi and insects disperse from nearby crop fields, causing root rot and herbivory. The list goes on.

Ginseng is our "canary in the coal mine" for what is happening to hundreds of other native herbaceous plant species in the eastern deciduous forest. Other than the harvest interaction with humans, ginseng is not particularly unique in the way it experiences the changes happening in the Anthropocene.

Can We Regulate Our Way out of Trouble?

The Endangered Species Act. In 1973, Congress, with overwhelming bipartisan support, passed the Endangered Species Act (ESA) to protect threatened and endangered species from extinction. Scientific studies of population trends of species listed in the ESA have clearly demonstrated the success of the act at both preventing extinction and improving prospects for recovery. Critics of the ESA point to the low level of delisting as evidence that regulations are not effective, but the science contradicts this. Recovery, especially for long-lived plants and animals, takes a long time, even when trends are in the right direction. In addition, since funds have always been limiting for endangered species protections, there was a tendency to list the most dire cases first, and these will naturally take longer to emerge to safe population sizes.

The ESA has not been an effective mechanism for protecting relatively abundant species that are in rapid decline, such as ginseng. For example, whitebark pines in the interior mountains of the northwestern United States and in western Canada were widespread and abundant across high elevations, but they have been decimated by fire suppression, blister rust, and mountain pine beetles. The recent bark beetle outbreaks have been exacerbated by climate change. The brown and dead trees stand out across the landscape as a stark reminder of the lethal effects of these stressors. Motivation to save these trees is strong since their seeds are a critical food source for species as diverse as Clark's nutcracker, red squirrel, and grizzly bear. Yet attempts to place whitebark pine on the endangered species list in order to achieve protections have failed repeatedly. The U.S. Fish & Wildlife Service (USFWS) stated in 2011: "After evaluating all the available scientific and commercial information regarding whitebark pine, the Service has determined protection under the ESA is warranted. However, listing whitebark pine at this time is precluded by the need to address other listings of higher priority." The agency went on to explain that funding was lacking and pointed out that there was little that they had the power to do to address the primary stressors, especially with respect to climate change. In other words, they threw up their hands. The ESA has protected endangered species with low population sizes and disappearing habitats much more effectively than widespread threatened species such as whitebark pine and ginseng. Or the innumerable other species being slowly extirpated by humans through lingchi.

Convention on International Trade in Endangered Species of Wild Fauna and Flora (CITES). Since ginseng is traded on the international market, it does qualify for regulation under CITES, administered by the USFWS. Because it is widespread and not "endangered" in most places, it is not listed in Appendix I, which is reserved for truly endangered species whose continued trade is a serious threat to cause extinction. Instead, ginseng was placed in Appendix II, a list including alligators, paddlefish, mahogany, and many corals.

Has CITES listing effectively regulated ginseng harvest and trade? The short answer is clearly no. The central evidence for this is figure 14.1, reiterated from chapter 4 (fig. 4.5), confirmed by model projections in chapter 5. It is important to note that the decade in which these data were gathered began nineteen years *after* the CITES listing of ginseng. Would ginseng have declined faster in the absence of the CITES listing? We do not know, but since the average decline for thirty populations was somewhere between 20 percent (optimistic scenario) and 30 percent (pessimistic scenario) over only one decade, we can conclude CITES is inadequate to protect wild ginseng. Population projections (chapter 5) confirm these numbers. Further corroborating this judgment of inadequacy is the high rate of noncompliance with regulations, in the form of

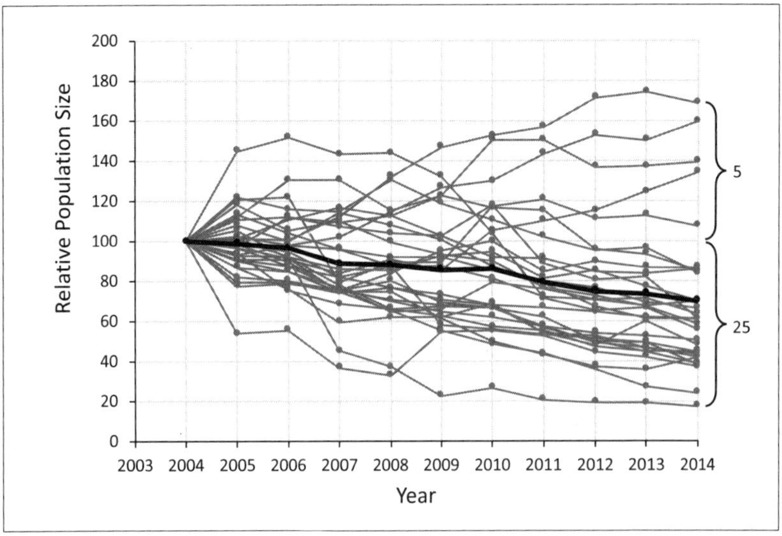

FIGURE 14.1.
Relative decline of twenty-five of thirty populations over one decade

TABLE 14.1
Agency managing each state's ginseng program

State	Agency Responsible for Ginseng Program
Alabama	Dept. of Agriculture and Industries
Arkansas	Dept. of Agriculture, State Plant Board
Georgia	Dept. of Natural Resources, Wildlife Resources Division
Illinois	Dept. of Natural Resources, Office of Law Enforcement
Indiana	Dept. of Natural Resources, Division of Nature Preserves
Iowa	Dept. of Natural Resources
Kentucky	Dept. of Agriculture
Maryland	Dept. of Agriculture, Plant Protection and Weed Management Section
Minnesota	Dept. of Natural Resources, Section of Wildlife
Missouri	Dept. of Conservation
New York	Dept. of Environmental Conservation, Division of Lands & Forests
North Carolina	Dept. of Agriculture, NC Plant Conservation Program
Ohio	Dept. of Natural Resources, Division of Wildlife
Pennsylvania	Dept. of Conservation & Natural Resources, Bureau of Forestry
Tennessee	Dept. of Environment and Conservation, Division of Natural Areas
Vermont	Dept. of Agriculture, Food & Markets, Plant Industry Section
Virginia	Dept. of Agriculture and Consumer Services
West Virginia	Division of Forestry
Wisconsin	Dept. of Natural Resources

(a) taking plants that are smaller than regulated size limits, (b) harvest of plants on land where permission has not been obtained or harvest is outlawed, and (c) illegal harvesting outside of harvest seasons.

As a federal agency with a very limited budget, the U.S. Fish & Wildlife Service devised a "ginseng program" that cleverly offloaded the burden of regulation on individual states that wanted to allow harvest. States placed their ginseng programs in a wide variety of agencies, most not well equipped to regulate a wild-harvested plant trade (see table 14.1), from departments of agriculture to departments of forestry, conservation, and wildlife. In an insightful review, Brent Bailey pointed out in his 1999 dissertation that each management paradigm had strengths and weaknesses. However, programs placed in state wildlife departments had distinct advantages in terms of management capabilities, including established permitting processes (licensing), potential for direct lines of communication with harvesters, and law enforcement. Only three of nineteen states placed their programs there. Lacking these capabilities in most states, ginseng regulation has been placed in agencies that often do the bare minimum to meet federal reporting requirements. I have met with and sympathize with the ginseng state coordinators; they are good folks who take their jobs seriously. But typically ginseng is only a minor, seasonal component of their assignment, and they are stretched thin trying to accomplish all their assigned tasks.

The Ginseng Curveball: What Is Wild?

Most species that are regulated for international trade are wild. Where there is a significant market, incentives will naturally arise for domesticating the animal or plant. The exceptions will be cases where domestication is difficult or impossible due to traits of the species that resist such efforts, or where markets specifically value traits only found in wild-raised individuals. Ginseng is both difficult to cultivate, making it a less than ideal crop, and valued wild for the root's gnarly, twisted character, only manifested by slow growth in natural, unamended soils.

All of this complicates efforts to regulate ginseng since CITES and the ESA are designed to protect wild native species in their natural habitats. Ginseng roots enter the market with several different origins: (a) truly wild: found in natural, untended woods and not planted or tended during growth by any direct human intervention; (b) wild-stewarded: seed planting within a wild population facilitated by humans and/or ginseng adults made less conspicuous to harvesters by clipping before or early in harvest season; (c) wild-simulated: planted, with seeds (often produced under cultivation method "e" below) sown on naturally or minimally disturbed soil (raking and replacing leaf litter is sometimes done) at relatively low density, mimicking natural populations; (d) woods-cultivated: seed (typically from cultivated sources, as in "e" below) sown at relatively high density in prepared beds under natural forest canopies; and (e) cultivated: grown under shade cloth or lath houses in prepared agricultural soil media. Both cultivated methods ("d" and "e") require significant inputs of fungicides and pesticides to prevent decimation of the crop prior to harvest.

For regulators, the difficulty is that the humans involved in methods "c" and "d" wish to obtain the maximum price for their roots as determined by the Asian traditional medicine market. Often prices for wild roots ("a" and "b") are tenfold (or more) higher than for cultivated roots, and prices for "c" and "d" roots fall somewhere between. Therefore, it behooves growers to have their roots treated as wild for the purposes of selling, though Asian buyers are typically savvy at recognizing the difference. If regulation of wild ginseng becomes too restrictive, their enterprise is threatened because they would need to comply with regulations. This conflict has motivated some growers to try to permanently blur the difference between wild and grown ginseng, claiming that there is no such thing as truly wild. Genetic analyses contradict that assertion, and, at least in Appalachia, where forests have always been extensive, the notion that hundreds of thousands of ginseng populations were planted is absurd.

The blend of root sources entering the "wild" market makes it difficult if not impossible for ginseng managers to use harvest totals to gauge the health of wild populations. Numerous rural advocacy groups have pushed wild-simulated and woods cultivation of ginseng as a secondary crop income for woodlot owners. Due to concerns about theft, woodlot owners rarely publicize their activities, and the number engaged in some form of these methods is unknown. On the face of it, this grown ginseng should benefit conservation of wild ginseng to the degree that the crop substitutes for wild. And it probably has, but until the evidence presented in this book was published, the woods-grown roots have masked what is happening to wild populations.

Environmental Regulations Can Work

Environmental regulations can be effective when they address the problem. For example, readers of a certain age will remember air pollution in most major U.S. metropolitan areas that reached dangerously high levels in the 1960s. Smog like that is experienced in Beijing today. But the Clean Air Act of 1970 allowed the states and EPA to set standards for automobile pollution resulting in vehicles that are 98–99 percent cleaner in terms of pollution-producing emissions than cars from the 1960s. Poisonous lead was removed from fuel. Air quality has improved despite ever increasing human populations. Standards were implemented. Innovation followed. The EPA estimates that for every dollar spent to reduce emissions, there were nine dollars in benefits to public health, worker productivity, and consumer savings.

In 1990, amendments to the Clean Air Act incentivized power plants to reduce sulfur and nitrogen oxides, significantly reducing acid rain. Lakes and streams in many areas of the northeastern United States have steadily gotten better, though we are not out of the woods yet.

Another resounding success story in environmental regulation is the Montreal Protocol of 1989, the first international agreement on regulation of chemical pollutants. This was made possible by, again, identifying and addressing the source of the problem. Chemicals called chlorofluorocarbons (CFCs), used widely as refrigerants, were working their way into the atmosphere. Once there, they destroyed the ozone layer. As it turns out, the ozone layer was acting as a shield, blocking high-energy-ultraviolet (UV) radiation from reaching the earth's surface. Loss of stratospheric ozone was a disaster in the making for human health worldwide. Fully 193 nations signed the agreement to phase out CFCs and rescue the ozone layer. The world we avoided by collectively signing on to this pact would have caused unprecedented rates of skin cancer, suppressed immune response, premature aging, and cataracts. To go outdoors would have meant slathering on sunscreen and wearing UV protective sunglasses on cloudy days. Studies showed that nonhuman animals and plants would have been severely affected as well.

Regulating Lingchi Effects

The difficulty with conserving species experiencing lingchi is obvious. Removing one kind of cut won't solve the problem. There are still many other cuts sufficient to cause decline.

Honeybees, a nonnative species in the United States but the producer of honey and an important pollinator for agricultural crops, provide another example of lingchi effects. In recent years, bee colonies have been dying at unprecedented rates in a phenomenon called Colony Collapse Disorder (CCD). Given the economic engine that honeybees drive, hundreds of scientists have been working hard to understand what is going on. The best explanation so far is that there is no one explanation. Instead, CCD is the result of multiple interacting factors. These include varroa mites, a variety of weed killers, pesticides, and fungicides used on crops in agriculture, and viral, fungal, and bacterial diseases made worse by the other factors.

So how do we solve a problem like honeybee die-off? Or ginseng decline? Or the Anthropocene decline of songbirds, the majority of slow-growing, native understory plants of the eastern deciduous forest, and more. Where do we go from here?

CHAPTER 15

Homo sapiens; Wise Human?

> The saddest aspect of life right now is that science gathers knowledge faster than society gathers wisdom.—**ISAAC ASIMOV**

We have reached a critical juncture for wild ginseng and its understory community friends. Knowing this, what will we do to ensure that ginseng is with us seven generations hence?

Scientists rarely make such policy decisions. Our role is to lay out possible futures, along with consequences, based on the best science available to us now, and then let others decide. At the end of this chapter, I will take off my scientist hat and express my personal opinion about what I think would be wise. But ultimately it will not be up to me. It will be up to you, the stakeholders in this drama.

Having just explained lingchi problems from a conservation perspective, I now address our direct interaction with ginseng—harvest—the factor in which ginseng differs from most other species experiencing lingchi. In chapter 16 I will address how we must respond to all the cuts we are inflicting.

Some would argue that governments should get out of the ginseng regulation business. This refrain is heard most loudly from those wishing to grow and profit from ginseng as some kind of crop. But wild ginseng (the noncrop) must be regulated as long as CITES exists, ginseng is listed on it, and we agree that unregulated international trade could negatively affect the wild populations (and there is no question it would, as long as a premium is paid for wild roots). Below I lay out options to be considered by federal and state regulators. But how harvesters respond to the future actions of state and federal regulators will determine whether we succeed in saving ginseng from the fate endured by its Asian cousin.

Option 1: Business as Usual

The sobering projections in figure 15.1 make some fundamental assumptions that are important to make clear. First, if we intend to generalize about the species as a whole, the sample of thirty populations used in the 2004–2014 monitoring is critical. We have never claimed that they are a perfectly random sample (see chapter 4). Statistically they do not qualify on that basis. But we do claim that they represent the range of natural populations occurring in the wild. So while the average rates of decline and increase may not give a perfect statistical mean, they can nevertheless be used as guide to what is happening.

Given the overall decline observed from 2004–2014, the two lines showing the future in figure 15.1 represent a "worst case" (solid line) and a "best case" (broken line) for persistence if nothing changes in our policies regulating harvest and nothing changes in any of the other stressors we have discussed. If, for example, harvest pressure, measured as number of plants taken from the wild, remains constant in the future, and this is the primary cause for decline, the straight line (extinction in about 2037) may be the more accurate projection. We already know that many negative pressures on ginseng are more likely to operate on a percentage basis; in other words, the "percent risk" of mortality might be high enough to cause decline, but this percent risk may not be changing. If the percent decline remains constant, and our 2004–2014 data are representative, the dashed line may be ginseng's future, all else being equal.

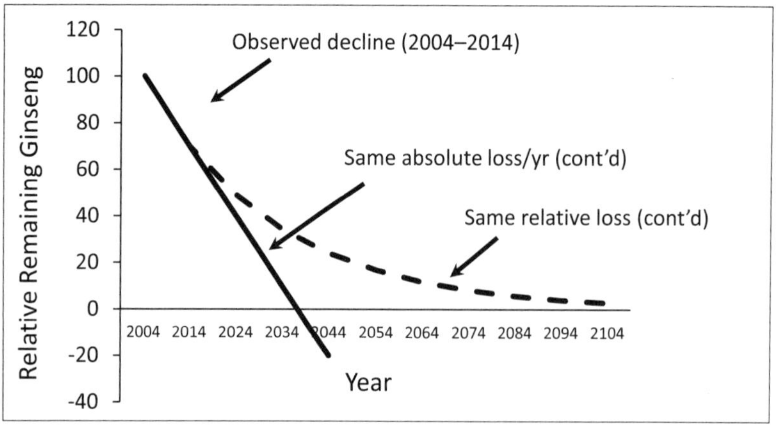

FIGURE 15.1.
Projection of observed rate of decline from 2004 to 2014 into the future. The solid line assumes the same absolute loss of plants per year, while the dashed line assumes the same relative (percent) loss as observed in that interval.

Again, both of the projections in figure 15.1 assume business as usual—that is, no change in harvest regulations, no worsening of climate change, continued overbrowsing by deer, et cetera. Both projections suggest a bleak future for ginseng if we continue on the present course.

That is the first option: watch ginseng rapidly or slowly become more and more rare and eventually become functionally extinct in the wild. Although the time frame may have been different, this is essentially the fate experienced by *Panax ginseng* (Asian ginseng).

Option 2: Tweak Current Regulations

To understand the second option, we need to understand what the current regulations are. I have compiled them, but they change in small ways every so often. So rather than produce a state-specific table here that will be out of date shortly after publication, I'll summarize the current regulations here in a general way, then we can consider what tweaking them might entail.

Only the basic parameters for regulations are laid out by the U.S. Fish & Wildlife Service. They have delegated the making of detailed ginseng-related laws to each state's regulatory agency, which gets approval from the state legislature. The states then require approval from the USFWS to go forward with a ginseng program. And, of course, in order for the program to be approved, states must convince the USFWS that ginseng will be adequately protected by their rules. This mechanism leads to broad coherence across different state programs with respect to a few elements but surprising diversity on others. For example, all states except one now have a harvest season start date of September 1, the lone exception being Illinois, which chose the first Saturday in September. Another common feature is that harvesters are required to plant ginseng seeds on-site and are not allowed to carry them off-site. All states require permission of the landowner for a harvester to dig on private property. In addition, with rare exceptions, digging on state lands (state parks or state forests) is prohibited.

The first obvious opportunity to tweak regulations is provided by states that diverge from the majority in a way that could aid conservation. I discuss these in turn:

1. **Licensing.** Four states require licenses for harvesters. The costs of these vary widely (from $2 in Maryland to $30.75 for nonresidents in Wisconsin). As with hunting or fishing licenses, requiring a license for ginseng harvest could have many benefits. First, and perhaps most important, it represents

an opportunity to educate harvesters about how to harvest in an ethical, stewarding manner. This is akin to state wildlife departments, for example, explaining in the literature on their licensing web pages, how to perform catch and release in a manner that fosters fish survival. Regulatory agencies would not have to reinvent material here; the American Herbal Products Association has developed effective brochures on this topic for each state already. Second, it could be a source of revenue for conservation, especially for the larger states with thousands of diggers. These funds could be used to perform ginseng monitoring in their states so they would not have to rely solely on harvest amounts to judge the status of wild ginseng. Third, if licenses were shown to dealers upon sale, it would allow state agencies to track characteristics of the harvester community, including important traits such as geographic distribution, number of harvesters, and root weight gathered per harvester.

2. **Size requirements.** With one exception, all states require that only plants with three leaves or more may be harvested. Illinois upped this to four leaves, although it is unclear whether the regulators understand that the vast majority of ten-year-old plants (97.6 percent, based on a sample of 545 ten-year-old plants) do *not* have four leaves. Nevertheless, there is variation in this requirement. Some states specify three leaves with five leaflets on each leaf. Based on analysis of nearly twelve thousand three-leaf plants measured in our thirty natural populations, this additional leaflet specification eliminates about 40 percent of three-leaf plants from eligibility for harvest (fig. 15.2). The reason for specifying five leaflets on each prong is to ensure that the plant is a fully developed "mature" adult. Indeed, seed production is 2.3-fold higher for the subset of three-leaf plants that have five leaflets per leaf, compared to smaller three-leaf plants (see fig. 15.2). In theory, the 555 leaflet requirement (five leaflets on each of three leaves) should improve the sustainability of harvest if seeds are planted on site.
3. **Age requirements.** The USFWS requires that ginseng plants be at least five years of age, and seven states explicitly adopted this requirement. The others appear to rely on the size requirement to ensure that plants are five years old or older. Two states (Vermont and Illinois), however, have adopted a stricter standard of ten years. Recall from chapter 5 that plants have not naturally replaced themselves until somewhere between twelve and twenty-three years of age, thus even the ten-year standard may not be stringent enough. Clearly a simple opportunity to improve regulations is to increase the age requirement. This has the additional advantage of being verifiable if rhizomes are left attached to the root.

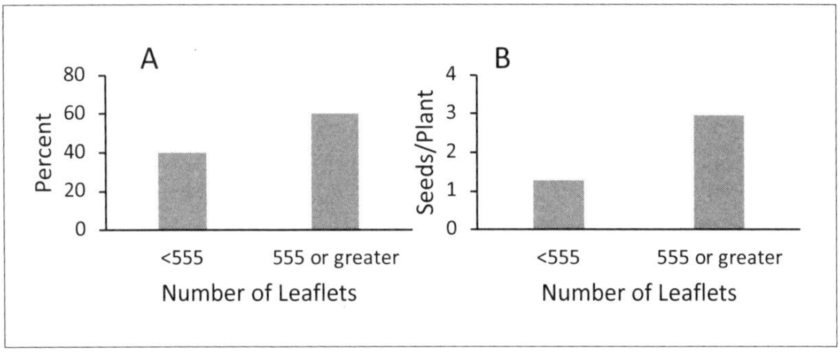

FIGURE 15.2.
(A) Percent of three-leaf plants having fewer than five leaflets on every leaf (<555>) versus having five or more leaflets per plant (555 or greater), and (B) mean reproductive output of these two three-leaf plant categories

4. **Fruit/seed requirements.** No states specify that a plant must have berries and seeds before it can be harvested. If such a requirement was implemented, the 60 percent of three-leaf plants that do not have seeds would not be eligible for harvest (N=11,917). Much more common are regulations specifying that plants cannot be harvested if they have green (unripe) berries. Others specify that berries must be red. None of the regulations seem to acknowledge the fact that the majority of adult plants have no berries.
5. **Prohibitions on harvest.** With minor exceptions, states have prohibited harvest on state lands. National forests are surprisingly variable, however. When harvesting is allowed, it is through a permitting process. Some national forests cap the number of permits and use a lottery system to determine who the lucky diggers are. About half of the national forests have outlawed digging, presumably out of concern for sustainability. The National Park Service uniformly prohibits harvesting.
6. **Harvest seasons.** (a) Onset. This is presently fairly uniform (only Illinois diverges from the rest in having a harvest season start of the first Saturday in September instead of on September 1). A minor tweak would be delaying harvest season until a later date, such as September 15. The advantages would be minor; although more seeds would be ripe by that point, the opportunity for diggers to plant them might be reduced since they might have been gravity-dispersed by that date. (b) Termination date. The end of the harvest season varies widely across states (from October 31 in Vermont to December 31 in Minnesota). This could be standardized, though again it would make little difference since most plants have senesced by late October and cannot be found by diggers.

7. **Summary.** Tweaking existing regulations to maximize conservation benefits could include (a) requiring states to have a licensing system comparable to that for fishing licenses, with both resident and nonresident licenses, (b) implementing a uniform minimum-size rule: three leaves, with five leaflets on each prong, with verification by turning in both stalks and roots to the dealer, (c) changing to a uniform ten-year age minimum for harvest, and (d) eliminating harvest on any national forest land.

If a harvest regulation is not verifiable, it can be easily subverted or ignored by dishonest harvesters (see chapter 9). Unverifiable regulations are essentially unenforceable, akin to having speed limit signs on a highway but giving law enforcement officers no equipment to measure speed.

The states with digger licenses can have dealers verify and record licenses at the point of sale. Currently only one of the other categories of existing regulations is verifiable: the one that requires harvested plants to be at least five or ten years old, verifiable by counting bud scars. Three states (Indiana, Missouri, Wisconsin) require diggers to carry tops (minus seeds) with them until they reach their place of residence; this allows verification of size compliance by conservation officers *if* the harvester is intercepted between the digging site and their residence. Since tops do not have to be turned in to dealers along with the roots, verification is currently not possible beyond the rare officer-digger encounter.

Option 3: Adopt New Regulations

Adopting new regulations tends to be more difficult than altering existing ones. A further challenge is that there are actually nineteen sets of regulations, one set for each state. This could also be used to advantage, however, as certain states could be used as incubators or testing grounds for the new rules, and they would not necessarily be adopted by all until proven effective. This would not work, however, unless the results were evaluated for conservation success. I foresee several possible new regulations to significantly enhance ginseng conservation:

1. **Mandatory digger training with license requirement.** Just as society has agreed that a driver's test must be passed before someone should be allowed behind the wheel of a car, it also seems reasonable in the face of much demonstrated reckless behavior by harvesters that training should be required to obtain a harvester's license. In selling any ginseng to a dealer, a valid harvester's license would have to be shown and recorded by the buyer.

Digger training could be funded by license fees and could be performed by the state ginseng coordinator in each state that permits digging, or by rigorous online testing.

2. **Science-driven age rule.** To be optimized, the age threshold for harvest should be tailored to the site; a good site might have a ten-year rule, while a poor site would *never* be harvestable without detrimental effects. However, site-by-site regulations are unrealistic. Using the guide of natural replacement as a function of age (chapter 5), a threshold between twelve and twenty-three years seems reasonable. To make it an easy number to remember, perhaps age fifteen should be the compromise threshold age. Alternatively, since aging rhizomes can be tricky, especially in the field, a minimum rhizome (neck) length could be specified that would ensure a minimum age threshold would be met (e.g., one inch of rhizome equals about fourteen years of age). This would have the added advantage that, on a good site, this requirement would be met at a younger age due to fast growth, while meeting this same length requirement at a poor site would appropriately require more years of growth. This requirement would also be verifiable at the point of sale.
3. **Fruits required.** This rule would reduce the number of legally harvestable plants but ensure that the opportunity existed for the harvester to plant at least one seed to replace the plant taken, an action that has been shown by simulations to improve the sustainability of harvesting (as shown in chapter 5, one seed is not enough, but it is better than zero). Verification would be challenging, but two mechanisms are possible: (a) if the harvester is required to turn in flower/fruiting stalks (minus berries), it is possible to determine whether the stalk bore fruit by examining the pedicels (tiny stalks under each flower/berry); a swollen pedicel means there was a berry developed on that flowering stalk, but if all of them are thin and undeveloped, there was no berry; (b) the dealer could require the harvester to show digital photos of each fruiting stalk. Since smartphones are becoming nearly ubiquitous, this requirement, at least as an alternative to mechanism "a" above, could work. Even if verification is difficult for this rule, simply stating the requirement educates the digger that leaving behind seeds is critically important.
4. **Location.** Eliminate all harvesting on public lands, including national forests. In this way, all harvest would be on private land. Require harvesters to show the written permission obtained for harvest (with landowner contact information) to the purchasing dealer. Dealers or ginseng coordinators could perform an audit of compliance by contacting a subset of landowners

to verify permission. Licenses could be revoked for diggers without proper permit, or fines could be levied akin to those associated with driving without a license.

5. **Bag limits.** Implement a maximum annual harvest per person (harvest limit). This is already done in at least one national forest. In this way, harvest could be capped to help preserve the resource. Such limits are routinely done for the hunting and fishing community, and the strong overlap of these groups with ginseng diggers would mean such limits would be inherently understood.
6. **Size criteria.** Implement a harvest threshold based on size *in addition to* age. The three-leaf, five-leaflet threshold could help in this regard, but there may be more effective size measures to use, such as stalk height. Retention of stalks after harvest would be needed for verification of either. More research is needed on how size restrictions could make harvest more sustainable. Such a requirement would be akin to minimum-size thresholds for fish that are common for certain species to ensure successful recruitment of new offspring.
7. **Fresh roots only.** Require diggers to sell roots immediately after harvest to dealers. Presently, diggers typically dry roots, though there is also a market for fresh roots and tops. If only fresh roots could be sold, this could make it easier to require that tops be brought in to verify size requirements, and it could make out-of-season harvest unprofitable and less likely to occur since roots cannot easily be kept fresh over many weeks. In addition, there is the potential for a market for fresh ginseng leaves; they contain ginsenosides too and could find a place in the market.
8. **Flexible harvest seasons.** In a drought year, to minimize further damage from fall harvest, the season could be suspended for the year or sharply shortened to permit wild populations to recover. Likewise, if wild harvest declined for the previous X number of years, the harvest season could be suspended or shortened until evidence for recovery was obtained.

Option 4: Preservation by Substitution: Shifting to Forest-Grown Ginseng

To an unknown degree, this transition from wild harvesting to wild-simulated or woods-cultivation (methods "c" and "d" from chapter 14) is already underway. Henceforth, I will refer to these two methods collectively as forest-grown and those who carry out these practices as forest growers. Because both wild harvesters and forest growers want maximum price for roots, roots from both sources are marketed as "wild" when possible. For a given volume of roots purchased as "wild," the greater the percentage of these that can come from forest-grown sources, the fewer truly wild roots will need to be harvested. Therefore, forest-grown approaches are to be encouraged.

At the same time, because sales of roots from forest growers are presently conflated with wild-harvested (roots collected using methods "a" or "b" in chapter 14) sales, and sales volume is one of the main statistics used by states and the USFWS to assess the status of wild populations, inferences about the true status of wild populations from these data are flawed. Only independent data on truly wild (not forest-grown) populations can rectify this situation. Therefore, it would benefit management of wild populations if the two harvest streams were formally separated. This is easier said than done since forest growers wish to receive the inflated prices paid for truly wild roots.

One creative but practical solution to this problem would be to separate the harvest streams by age. First, the present five-year age minimum rule would be eliminated, with harvest of truly wild populations restricted to roots age fifteen and older (see option 3, item 2, above). By contrast, forest-grown roots would be required to be no older than ten years of age, thus achieving the desired separation. Since the Asian traditional medicine market values older roots, a price separation would also occur, but forest-grown roots would still retain much higher value than shade-cloth-farmed roots. This separation could be accomplished with little sacrifice on the part of forest growers, since most such operations plan to harvest on a cycle shorter than ten years anyway; longer harvest cycles risk losing the crop to disease or theft or both.

Presently, forest growing (using wild-simulated or woods-cultivated growth methods) is impeded by some state and federal regulations. This is unfortunate because it prevents the substitution of forest-grown ginseng for truly wild ginseng that would effectively conserve the latter. To encourage wild-grown ginseng operations, ginseng sold into this market should be absolved from CITES regulations (after all, CITES is intended to conserve wild populations of a species). This would mean that forest growers could harvest plants of any size or age (up to age ten), could harvest at any time of year, and be exempt from on-site seed planting requirements.

Along with the new freedoms to pursue forest "farming" of wild-grown ginseng would be a legal responsibility of each grower to certify the growing operation. This would ensure that roots harvested from wild populations are not being sold as forest-grown. Retired Cornell extension agent and ginseng forest grower Bob Beyfuss suggests three-party (not third party) verification before roots could be sold as forest-grown. He lists possible certifiers of forest-grown operations: (a) agricultural extension agents, (b) Farm Service Agency employees, (c) local law enforcement officials (police, deputy sheriffs, game wardens), or (d) other state agency employees (foresters, conservation officers, etc.). These individuals could perform site visits and request records of seed purchases, evidence of attendance at workshops, and recordkeeping of planting and monitoring. A standard form, signed by three such parties, would be difficult to fake.

One impediment to nurturing a forest-grown ginseng industry is a lack of local seed sources. Currently seeds from local wild populations must be planted on-site, a rule intended to ensure regeneration from within. However, this prevents would-be seed purveyors from legally acquiring local genotypes to sell to forest growers. A state's ginseng program could issue limited permits for seed and live plant collection for the specific purpose of developing local seed sources for growers, but this would require sanctioning by the USFWS and a clearly written protocol so as not to damage wild populations. Ultimately, developing local seed sources would have several benefits for local forest growers, including (1) greater success of plantings owing to use of genotypes adapted to the local environmental conditions and (2) enhancing niche marketing opportunities associated with particular genotypes. This is precisely what is happening in the rapidly expanding CBD and cannabis trade after legalization. Branding, either by genotype or location of origin, provides added value to products, particularly when backed by scientific evidence for differences in biochemical and medicinal properties from other products. Think "Vermont maple syrup." "Kentucky forest-grown ginseng" could gain cachet among consumers similar to "Kentucky bourbon."

Option 5: Ban Harvest Now

The Endangered Species Act allows listing of species under two categories: endangered or threatened. An endangered species "is in danger of extinction throughout all or a significant portion of its range." A threatened species is one that is likely to become endangered in the foreseeable future. Ginseng is clearly not endangered yet, given the volume of annual harvest that occurs (with the caveat above that there is an unknown amount of forest-grown ginseng mixed with truly wild). However, figure 15.1 makes a projection into the foreseeable future, and it shows a species that is clearly threatened. As the foregoing chapters have demonstrated, it is threatened not just by harvest but also by a multitude of other factors acting in concert.

An alternative mechanism that would be less protective than ESA listing would be transferring wild ginseng from CITES Appendix II to Appendix I. According the CITES website, Appendix I contains species that are "the most endangered." They are "threatened with extinction and CITES prohibits international trade in specimens of these species except when the purpose of the import is not commercial."

Note that under existing laws there is a partial ban on harvesting now, on specific state lands and in virtually all nature preserves, all national parks, and some national forests. In other words, some land management entities have already determined that ginseng is threatened to the point where banning harvest makes sense. The effectiveness of any ban on harvest will rely on strict law enforcement.

IMHO

With respect to harvest, changes from the status quo are needed, so Option 1 above is unacceptable. Elements of Option 2 would be the easiest changes to accomplish because some states have already adopted them, so a model exists for statutes that could be adopted by other states. Digger licensing might be one of the most challenging elements to implement but could also be one of the most effective. Implementing a science-driven age rule could help greatly, although it would be most effective if implemented alongside separate rules for forest-grown ginseng to partition the harvest streams into clearly definable parts (wild vs. forest-grown). Requiring that plants have fruits would be difficult to verify, so I think this would be most effective as a strong recommendation included with educational materials that all licensed diggers would receive. The inconsistent policy among national forests can cause confusion, so I would advocate for banning harvest on all public lands. In this way, harvest would have to occur on private property only, and every sale to a dealer should require a written, signed permit from the private landowner. Biologically, a size threshold makes sense, but implementing a size requirement that is not verifiable will be nearly useless. At present we need better research on the consequence of setting various possible size thresholds prior to implementing a size rule. Whatever criterion is chosen should be verifiable at all points, both in the field and at the point of sale. With respect to adoption of new regulations (Option 3), all of these could help preserve ginseng. Given the urgent need for better harvest regulations, these restrictions could help drive a shift to forest-grown ginseng to satisfy the market demand for gnarly, twisted roots. Option 4 could be the ultimate tool to remove harvest as one of the top-three stressors of remaining wild ginseng populations. Therefore, forest growing should be encouraged by state and federal regulations. Further research on optimizing wild-simulated and woods-cultivated growth methods is urgently needed, and state universities within ginseng's range should be supporting these efforts (this is already happening in a few places). To reiterate, developing local seed sources is to be encouraged by whatever means possible, to (1) ensure greater success rates, (2) develop branding opportunities, (3) provide another business opportunity involving ginseng, and (4) prevent genetic contamination of wild populations with cultivated seed sources. Option 5 could be considered a last resort only if Options 2–4 do not succeed. Adoption of Option 5 (either through CITES or the ESA) seems almost inevitable if current trends continue.

CHAPTER 16

Toward a New Deal with Nature

> Governments and businesses now have an opportunity to take a critical, collective step to arrest this [biodiversity] decline: to agree to protect at least 30 percent of the world on land and sea.—**THE NATURE CONSERVANCY**

So far I have laid out options with respect to how humans interact with ginseng in only one dimension, harvest. But the premise of this entire book has been that harvest is not ginseng's only problem—that, in fact, wild ginseng is subject to lingchi, death by a thousand cuts, not death by harvest alone. Therefore, regardless of the future of human harvest of wild ginseng, none of the options developed in chapter 15 may be sufficient to rescue ginseng from its current path toward extinction.

The Danger of Pessimism

A pessimist's response to this bind we find ourselves in would be to simply throw up one's hands and give up trying to save wild ginseng and indeed all of the wild native herbaceous plants that grace our diverse forest understories. A pessimist would take the stance that if improving harvest regulations is not going to save the species, let's just continue to exploit it until it's gone.

Singer-songwriter Joni Mitchell wrote and sang "You don't know what you've got till it's gone." But in the case of ginseng, though we are learning more all the time about its medicinal properties, and we have much more to learn in every dimension, we do actually have a pretty good idea what we've got. It's America's most treasured wild medicinal plant. And if we add all the other forest understory plants to the mix, we have an even better idea what we've got. We have potential medicines for the most serious diseases afflicting humankind. We have the base of the food chain supporting a diverse community of insects, amphibians, reptiles, birds, and mammals. We have beautiful leaves and flowers and fruits and seeds that enrich every walk in the woods. We have both quantifiable and unquantifiable wealth. We have this great gift, bequeathed to us by nature, that we could protect for our children and grandchildren. Or we could lose it.

Why Optimism Is Better

The optimist cannot abide the notion of simply letting it go, of leaving an impoverished world to our children and grandchildren in which we have squandered the wealth found in nature, what I will refer to as its biotreasury.

I have always been an optimist. I believe truth wins out in the end. After much gnashing of teeth, Rachel Carson's clarion call about the dire effects of DDT ultimately resulted in its elimination from our mosquito-fighting arsenal and recovery of diverse raptors such as the peregrine falcon and the bald eagle and fish-eaters such as the brown pelican and osprey. Observations of an invisible, thinning ozone layer, along with the beginnings of a skin cancer epidemic, led to a nearly worldwide ban on chlorofluorocarbons. As pollution was shown to affect human health and destroy vibrant freshwater fisheries, the Clean Air Act was passed. Though more needs to be done, lakes and streams once devoid of life have begun to recover. The Endangered Species Act has saved many species in the United States from sure extinction. This is progress we collectively have made by using the power of self-governance laid out in our Constitution. Fundamentally, these are the achievements of optimists who believed we could be better.

The Path Forward on Climate Change

Our species, *Homo sapiens*, is struggling mightily to deal with climate change because the root cause is our old, entrenched energy production systems (based on coal, oil, and gas). These powerful industries have purchased influence in the halls of power. I have little doubt the collective "we" will solve this problem as well, because we will ultimately have no choice. But I am less confident about a peaceful, untroubled path to that solution. Because of our collective shortsightedness, combined with corruption of the public will, I worry that the best hope we have for finally tackling climate change is an unmitigated and hyperexpensive disaster that will also sacrifice an astronomical number of human lives. Probably the sooner, the better, for our own sake. Though our comeuppance could take many forms, I suspect it will most likely be in the form of a climate-change-fueled, extra-intense hurricane making landfall at one or more large cities at high tide, after the sea level has risen a foot or two more due to glacial ice melt.

Believe me, I hope I am wrong. I hope we recognize what lack of foresight will do. The climate scientists among us know what is coming. They have already seen that the light coming toward us in the tunnel is a freight train, not daylight. But Hurricane Katrina was not enough. Harvey was not enough. Those storms cost a mere $125 billion each, and the lives lost were a mere blip on the radar of most Americans. Apparently, the damage must be far worse for society to finally acknowledge that the long-term costs of doing little to nothing about climate change are far greater than the costs of addressing the problem now. It's a matter of time. Unless we the people begin taking the long view of our own welfare, elect politicians who cannot be bought, and take actions on our own initiative to live sustainably on the planet. Will we? How wise are we? Will we earn our species name, *sapiens*? The cost of shortsightedness in human lives and dollars could be immense. Climate change will teach us the value of listening to experts who know more than we do—and heeding their warnings.

Witness the coronavirus pandemic that broke out in 2020. The death toll and economic cost of ignoring what scientists have said for decades about a "coming plague" is hitting the United States and the world very hard. But wouldn't spending a few billion dollars on preventative measures be a smarter strategy? Shouldn't the world insist that humans stop exploiting certain wild species of bats, pangolins, and monkeys that are common sources of deadly viruses? In the United States, the cost is mounting into the trillions of dollars to address an issue, after the fact, that was preventable—if only we had listened to and funded scientists. Why do we not insure ourselves against the painful reality of horrendous climate change costs by taking preventative measures now? Would we prefer to sacrifice Miami, Fort Lauderdale, West Palm Beach, and Jacksonville at some time in the not so distant future? Could coronavirus be a great teacher after all, prepping us mentally for the greater storm to come?

The one-cause environmental problems (DDT, CFCs, fossil fuel use), as difficult as they can be for society to address, are solvable through substitution. Find another pesticide to kill mosquitoes to prevent malaria. Find another refrigerant that does not deplete the ozone layer. Find and use alternative energy sources that do not emit greenhouse gases and thereby avoid drastic changes to the climate.

The Path Forward on Biodiversity Loss

Biodiversity loss—that is, the extinction of species—is doubly challenging because it is both a slow-motion disaster (like climate change) and has no one root cause. Myriad environmental cuts are knifing into our rich natural heritage. Focusing on just one threatened species—ginseng—as we have in this book, illustrates this problem perfectly. The lingchi for ginseng—harvest, deer browse, climate change, invasive species, land clearing, logging, suburban sprawl, mining—all contribute in their own diffuse but persistent way, different in magnitude in each of thousands of populations but collectively adding up to net population decline and loss. I contend that thousands of other species we care about, but that have not received such careful attention as ginseng, are in the same boat. The problem is far larger than ginseng.

So, how do we solve a problem that does not have one victim, one cause, or one solution? Ginseng shows us that managing on a species-by-species basis does not work. We cannot manage ginseng harvest, for example, without understanding all the other stressors acting on populations. Harvest effects depend on the level of deer browse and in the future will depend on climate change, for example. We cannot manage deer in isolation without destructive consequences for the ecosystem of which they are just a part. We clearly need a broader vision. Our management paradigm needs to change. Our very relationship to wild species needs to shift now. We must view the ecosystem as a whole, and ecology tells us we must work to reestablish the natural processes that support biodiversity.

As we have assembled our ginseng-centric view of the world, we see more clearly what wild ginseng needs in order to survive and thrive. No harvest (or, at the very least, stewardship), with substitution by forest-grown crops. Sharply reduced deer browsing. Lower competition from invasive species. Mature forest canopies with natural gaps. Moist, well-developed soil profiles. A thriving community of mutualist pollinators and seed dispersers. A climate that remains close to the adaptive peaks for temperature and precipitation. In other words, ginseng needs mature forest habitat with a full complement of species that ensure its success. How can we give ginseng and other wild species what they need to survive and thrive in a human-dominated world?

Preserved Habitat Is Central

A fundamental requirement for every species is quality habitat. Indeed, the key to biodiversity preservation, as the enormous body of conservation research has borne out, is habitat restoration and preservation. Despite our four-hundred-year history of habitat destruction and degradation in the eastern deciduous forest, we find ourselves at a critical juncture where habitat for wild species can actually be conserved and expanded. Now, a century after the last great logging

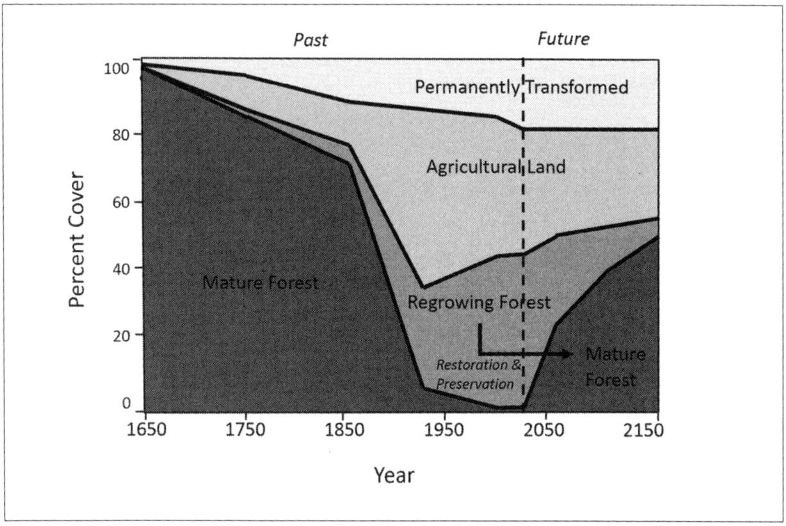

FIGURE 16.1.
Past history and possible future trajectory for the eastern deciduous forest, recognizing the critical role for biodiversity conservation in restoring and preserving mature forests

phase, and with a century of abandoned farmland returning to regrowing forest, the eastern deciduous forest of North America could be poised to recover and sustain biodiversity on a massive scale. I ask you to optimistically reimagine a bright biodiversity future based on figure 10.7, which outlined the historical pattern of land use in the eastern United States.

Imagine a future in which a large portion of regrowing forests are deliberately allowed to become rich, mature forests (see fig. 16.1). Through natural succession, these communities would become robust storehouses for biodiversity in ecosystems that function in balance from the top predators down to the smallest plants and even-smaller microbes. Natural disturbances would still occur in such forests, and the landscape would be diverse. But spaces would be found containing gigantic trees the likes of which have not been seen since the late 1800s. The preserved spaces would be large enough to allow top predators requiring large home ranges to reach viable population sizes. Herbivores would be kept in check.

The configuration of these preserved lands will be critical. Preserving north–south corridors and elevational gradients will be a key to preservation on a planet undergoing climate change. This will allow animals and plants to track optimal climates for survival. This idea is not so far-fetched as it might sound. In western North America, the Yellowstone to Yukon Conservation

Initiative is already successfully working to connect a two-thousand-mile-long gradient of wild habitats arrayed from north to south to ensure conservation of intact wild ecosystems. In the East, a more modest Algonquin to Adirondacks initiative is underway, but a much more ambitious effort is needed to connect Appalachian natural habitats. Only the initial pieces of this conservation puzzle are in place. Widespread, but uncommon to rare, declining species like ginseng will benefit enormously from increased protected habitat.

Biologist E. O. Wilson wrote in his 1984 book *Biophilia*, "The one process now going on that will take millions of years to correct is the loss of genetic and species diversity by the destruction of natural habitats." Wilson proposed a concrete solution: allocating half the earth (both terrestrial and ocean) for nature, ceding the rest to human dominance. (See the website of the Half-Earth Project: https://www.half-earthproject.org.) More ambitious, and longer-term in scope than the "30x30" plan of the Nature Conservancy (30 percent preserved by 2030), the half-earth solution provides a clear goal to work toward. The future depicted in figure 16.1 for the eastern deciduous forest in the United States is consistent with this goal, and it is achievable through deliberate action by individuals, organizations, and local, state, and federal governments.

The governmental mechanisms for preserving more land for our planetary companion species are already in place in the U.S. national parks, "America's Greatest Idea." Their protections and biodiversity-first policies must be expanded greatly and provide the core strategy for biodiversity preservation. Don't think we need more national parks? Great Smoky Mountains National Park tops the annual park visitation rate every year with more than eleven million visitors to a relatively small land area. The park is a vital component of the local economies, but it is in danger of being loved to death. The park rangers and staff are doing a superb job preserving this tiny gem of Appalachian biodiversity, but there are not nearly enough national parks to satisfy the deep urge people feel to be in nature, and to know there are wild places to visit, especially in the eastern United States.

The national parks solution to habitat expansion is ideal in several respects. First, people are vital components of national parks. Visitation is encouraged and managed at the same time. Nature education is a primary mission of the parks, a critical element for a human population that is increasingly disconnected from nature. Parks reestablish that connection. U.S. national parks cost a mere $3 billion annually, less than ten dollars per person.* Yet they return *ten*

*Don't think we have the money? Consider how easily the U.S. government allocated one thousand times this much to battle novel coronavirus in 2020, without batting an eye! Where there is a will, there are dollars.

times this amount to local economies, employing over three hundred thousand people. As part of a New Deal with Nature, a greatly expanded national park system would be an investment that would provide even greater returns to the national biotreasury.

Going forward, imagine a system of national parks preserving the highest-quality forests and landscape features, connected by national forests, state forests, and state parks and spanning the length of the Appalachian Mountains, from the Chattahoochee National Forest, Nantahela National Forest, Great Smoky Mountain National Park, Pisgah National Forest, and Cherokee National Forest in the South, to the Daniel Boone, Jefferson, and Monongahela National Forests in central Appalachia, and north to the Allegheny National Forest, the Catskills and Adirondacks, the Green Mountains of Vermont, and the White Mountains of New Hampshire. This is a vast land area. Imagine a maturing, regrowing forest across the entire length of the Appalachian chain, protecting ginseng and thousands of plants and animals unique to the region. A continuous corridor combined with plenty of elevational gradients would allow migration in response to climate change and sufficient land area for a full complement of trophic levels to restore balance to the ecosystem.

More difficult to achieve, but not impossible in the long run, would be a midwestern corridor running from the Ouichita, Ozark, and Mark Twain National Forests in Arkansas and Missouri north to the Chequamegon-Nicolet and Superior National Forests of the Great Lakes region. The farmlands of Illinois and Iowa, long ago converted to agriculture, interrupt this gradient, but elements of a corridor along and near the Mississippi River are already established in a series of wooded parks and preserves. These could serve as a basis for building a larger unit. Preserving this long connector would have the added benefit of preserving the vital floodplain areas of the river.

Complementing the national park system are national wildlife refuges to preserve key habitats for sets of species with specialized requirements. Now there are 583 refuges but precious few in the eastern hills and mountains. Wherever possible these need to be expanded to connect to other preserved habitats. Presently aimed primarily at animal conservation (e.g., waterfowl), wildlife refuges could be redesignated "ecosystem refuges," and more could then be created to include special unique habitats for plants, such as serpentine grasslands, shale barrens, or sphagnum bogs, all containing endemic species found nowhere else.

At intervals along the gradients of wild lands, we should recognize special areas to designate as wilderness, places where nature is left even more undisturbed by human activities. You may not know that there are already 803 designated wilderness areas in the United States, but only a small minority of these are found in the East. As forests mature and become "rewilded" in coming decades, more land area will qualify.

The easiest route to preservation might seem to be presidential decree. Unfortunately, national monuments, while easy for presidents to create, appear all too easy to undo by political whim (witness the recent history with Bears Ears and Grand Staircase-Escalante National Monuments in Utah). Of the 128 national monuments, most are in the West, and the eastern ones tend to preserve historic sites.

National forest and Bureau of Land Management lands tend to allow human exploitation to a degree that can threaten biodiversity. Therefore, these need more formal protections by redesignation of portions as national parks, wildlife refuges, and wilderness areas.

In the eastern United States, private land, vast tracts owned by lumber companies and mining companies, should be added to the national biotreasury when possible as well. For private landowners, strong incentives should be devised by state and federal governments to preserve their properties, either by donation (perhaps stimulated by tax incentives) or with conservation easements designed to save biodiversity.

All This for Ginseng?

I am fully aware that this grand vision for a New Deal with Nature may seem like a ludicrous conclusion to draw from a set of extensive studies on American ginseng, whether it is America's most valuable medicinal plant or not. But what has struck me with utter clarity as I have written about ginseng, is that this is not just about ginseng. It is about trillium. It is about hemlock. It is about leopard frogs, slimy salamanders, and mudpuppies. It is about box turtles. Wood turtles. Indigo snakes. Hognose snakes. Cerulean warblers. Wood thrushes. The list could fill an entire book unto itself. The bottom line is that it is about our national and global biotreasury. For four centuries, our biological wealth has been gradually eroding before our eyes. The time has come to reverse this shrinkage and begin saving. Indeed, to begin investing.

Rebuilding the biotreasury will require time and sustained planning, backed by public support. The vision is intended to extend seven generations. We are not used to thinking about the long future. And, in the end, solving the problem of extinction may well be more complex than fixing the ozone layer, stopping acid rain, or even halting and possibly reversing climate change. But it starts with giving our native species a chance, including habitat restoration, and adding substantially and methodically to the protected habitat set aside for their proliferation.

Worth remembering is this fundamental principle of population biology—*all species have the potential to grow exponentially in number*—even those, like ginseng, with slow life histories. Recovery can be rapid once habitat becomes available and protections are in place.

While recovery is underway, those placed in charge of our biotreasury must provide feedback to the general public about the fruits of our collective conservation efforts. Every species removed from the endangered species list because it has reached secure numbers and range should be celebrated with a one-time national holiday. Every example of a threatened species rescued from decline should be taught to schoolchildren as part of their science curriculum. Encouragement along a path to a restored relationship to nature requires hope and awareness, backed by data.

Knowing what we know about just one species in our diverse flora, ginseng, we are equipped with the lenses to foresee alternative futures with greater clarity. Our own ability to recognize and reverse our selfish, shortsighted, and destructive tendencies and replace them with a future of benign stewardship of the natural world will speak volumes about the improvement of our vision, about our own ability to be farsighted. Come to think of it, isn't being farsighted the very definition of wisdom, which should be a primary trait of *Homo sapiens*?

The need to adopt a New Deal with Nature is urgent. If not for ourselves, then for our children, our grandchildren, and beyond. Ginseng teaches us the value of taking the long view, in which rewards endure across generations. The story of ginseng offers us the opportunity to rewrite the story of our relationship to wild species.

Eleanor and Artie

August 16, 2030. "I don't know about this, Mom. Are you sure Artie is ready to go off in the woods with you? He's only ten. What if he twists an ankle or worse, and how would you get him out?"

"I was ten when Papaw took me out. He'll be fine. Besides, I'll have my cell phone in case of a problem."

Eleanor retired ten years ago, after thirty-five years as a nurse at the Charleston Area Medical Center. Throughout her career, she was known for her kind heart and soft manner with every single patient she encountered. It's the West Virginia way. But the coronavirus pandemic of 2020 nearly did her in.

She cared for patient after patient as they poured into her unit at the hospital, watching the slow-motion tragedy unfold as several senior patients under her care, and even a few younger ones, died lonely deaths of respiratory failure in coronavirus isolation units. Nationwide, over Memorial Day weekend, the death toll rose above 100,000. The number of deaths would surpass 800,000 in December 2021. Nothing else was in the news for months.

In June 2020, she came down with COVID-19 herself and spent three weeks back at the home place in McDowell County in isolation, cared for by her only daughter. She recovered, but the joy in her job was not what it used to be. Though she never said so, she was quietly angry that the politicians had not listened to the scientists, at least not at first and not before it was too late to prevent the carnage. Ignorance and denial at the top had delayed social distancing among citizens and caused a shortage of personal protective equipment for nurses and an inadequate supply of respirators for hospitals. Fearing for the economy, governors across the country had cancelled social distancing mandates too soon, and a second wave had struck in the summer that year, followed by a third devastating wave in the fall and winter after that. Science denialism was no way to run a state, a country, or for that matter, the world. The stark consequences were right in front of her eyes.

That visit home awakened in Eleanor a deep-seated desire to pivot toward what was truly important in life. Her family. Ellie's daughter had just brought a grandson into her world and named him Arthur, Artie for short. Ten years later, he was an eager, energetic, growing boy who sparked joy in every room he entered.

Artie bounded up the steps of his grandma's house, ready to go at 6:00 a.m. "Be careful out there!" Artie's mother yelled. The excited pair, grandmother and grandson, jumped into Eleanor's pickup and headed down the road to her brother's place. When they arrived, they parked down the road from the house.

Eleanor dropped her pack over the fence with a loud *clank* and held open the barbed wire for Artie to climb through. Then she followed, balancing gingerly on the lowest wire and climbing over. "What's in your pack, Grammaw?"

"You'll see," said Eleanor.

Off they trudged up the hill toward the forested hollow on the far side of the pasture. "What's that bird with the dark V on his chest sitting on the post, Grammaw?" said Artie.

"He's the one singing to us this morning, Artie. That's a meadowlark."

"I've never seen one before. He's perty!" She didn't have the heart to tell him that two-thirds of the meadowlarks in North America had disappeared since she was a child.

"Yup. He's a special one."

As they approached the woods, another avian friend was sending out melodic pairs of notes, each pair different from the last. As they stopped to listen, Artie made soft *swsh, swsh, swsh* noises. Moments later, a gorgeous tiny royal blue bird popped out of a bush not ten feet in front of them, eyed them quizzically, and just as quickly disappeared back into the forest.

"Indigo bunting!" Ellie beamed. "How did you know how to call him up?"

"My mom taught me that in our backyard."

"Wow, you are quite the naturalist, young man! Let's go see if we can find some ginseng."

The summer had been wetter than normal, and, as they made their way up the little stream, pools of water remained, with a small trickle joining them. Each pool was teeming with critters designed to distract a ten-year-old from the day's mission, but Ellie was patient. She was delighted he loved exploring the stream as much as she had as a child.

Finally, after identifying water striders, crayfish, and, yes, slimy salamanders, they reached the massive white oak tree that Eleanor had been looking for. "We are here!" she proclaimed.

"Where are we?" said Artie.

"My ginseng garden."

Shortly after she retired, Ellie met a man selling locally sourced ginseng seeds at the farmer's market. She said to him, with a smile, "That looks like my retirement fund!" He laughed and said "Could be!" She purchased two pounds and carefully planted them on her brother's back woodlot in a natural way she had read about online. The idea was to produce valuable "forest-grown" ginseng roots and thereby help to conserve the truly wild ginseng. Ten years later, here they were, ready to harvest.

"What is ginseng good for, Grammaw?"

"More like, 'What *isn't* it good for?' according to some folks, Artie. Mostly it's used to give people more energy and stamina. But millions of people think it can do much more."

"Do you take it, Grammaw?"

"As a matter of fact, I do. I put a little in my tea every morning."

"Maybe that's why you hike uphill faster than anybody I know, except me, of course. What do we do now?"

"This ginseng is ready to harvest, Artie. And you're going to help me, if you think you are up to it." She demonstrated how to carefully clear away the leaf litter around the stalk, and prod the "neck" with her finger to see which direction it went, echoing the lessons she had learned long ago. She explained how the neck would eventually lead to the root. Then she pulled an odd-looking tool out of her rucksack.

"What's that?" Artie asked. He had never seen anything like it.

"That is your great-great-grandfather's 'sang hoe. Watch. I loosen the soil all around the root, then pull away the clods with my fingers so I don't damage it." She worked her way around the stalk for several minutes, being careful not to nick the root. Finally, she pulled out an old screwdriver, and pried loose the last little rootlets to free the plant.

"Cool! That's an awesome looking root, and it sort of looks like a man with three legs! Can I try to dig one?"

"Sure thing. Let me watch you do one, then if you are careful enough, you can dig as many as you want." Turns out, Artie was a precocious ginseng digger. After just a little guidance from his grandmother, he was getting about ten roots in an hour. By the day's end, they were both worn out from so much digging, but Eleanor's sack was full. They made their way back across the field to her truck as the sun was setting, and she placed her arm around his shoulder, smiled, and said "I'm proud of you, Artie. With a little more practice, you could be a ginseng Jedi!" Artie just smiled back.

Artie and his grandmother would remember that day for the rest of their lives. Ellie returned to her plot several more times that fall to harvest. Though she had joked about stoking her retirement fund, she actually had something else in mind for the profit she made from her forest-grown ginseng garden that year. She donated every last bit of it to her grandson's college fund. He only learned about this years later when his mother told him he could go to any college he wanted to after graduating from high school.

"I don't get it," Artie said. "How can we afford it?"

"She didn't want me to tell you, but remember Grammaw's ginseng plot? Well—"

"Say no more." Artie knew.

Moments later, he rang up Grammaw Ellie on his cell phone. His voice cracked just a little as he simply said, "Thank you. You . . . are . . . the best."

CHAPTER 17

Notes of Optimism

June 8, 2020. Jen and I revisited Population 30, the "best population," for one last spring census. High temperatures in the city that day reached the mid-eighties, but under the tall canopy of diverse large trees, cool spring conditions prevailed. When we arrived, the temperature was in the low fifties, with a high by midafternoon of about seventy. Upon arrival, we were greeted by the twittering song of the winter wren, a chorus of warblers, and the haunting calls of not one but two of ginseng's mutualist seed dispersers, the veery and the wood thrush. The previous summer, a huge cherry tree in the center of the population had fallen, opening another gap in the canopy where additional light flooded in, stimulating growth of the already-lush understory.

Ginseng in this ideal environment is not without stress. Something (mice?) had chewed on some of the stalks, while others were mysteriously missing their inflorescences. Still others had been browsed by deer, at a slightly higher rate than prior years. A fungus appeared to infect a portion of the plants. But we could sense something was missing from the environment of this healthy ginseng population. That something was the dominant effects of direct and indirect stressors wrought by us, *Homo sapiens*. I am sure, if we looked hard enough, we could find them. CO_2 levels were higher than normal. Soil pH was lower than normal due to acid rain. Pollutants may have drifted in from faraway cities. But by and large, this was a population that had been given a chance to be itself; individuals within it could grow large, live long lives, produce copious crimson fruits and seeds, and die of natural causes. In the twenty years we visited this special place, we saw no harvest, moderate deer browse, few invasive species, diverse native species, no timber cutting, no mining, no housing development, and only natural disturbances.

The census took an inordinately long time to complete, for a wonderful reason. We added pages of new entries into the books. Dozens of entries like this: "Plant ID: 1123, 1 prong, 3 leaflet. Location: 38 deg, 10 cm from #286. NS [new seedling]." The population was recruiting the next generation.

As we walked out of the woods that day, with the light fading, one last wood thrush call rang out, perfectly capturing the exuberance I felt, knowing that healthy ginseng populations existed somewhere, "out there," in the deep woods. Populations containing possible cures for human disease. Or not. But populations that were a key link in a healthy ecosystem and, indeed, a healthy world. Definitely. *Bup bup, EEE-O-LAY!* It was the sound of hope for ginseng. And us.

CHAPTER 18

What Can You Do?

> The Peacemaker taught us about the Seven Generations. He said, when you sit in council for the welfare of the people, you must not think of yourself or of your family, not even of your generation. He said, make your decisions on behalf of the seven generations coming, so that they may enjoy what you have today.—**OREN LYONS**

Knowing that ginseng is under threat from many quarters, it is appropriate to end this book by answering the question that will be first and foremost on any sympathetic reader's mind: "What can I do to save wild American ginseng?"

Message to Private Landowners

Private land dominates the landscape across the eastern United States; therefore, what private landowners do is absolutely key for long-term ginseng conservation. Regardless of how the U.S. Fish & Wildlife Service and the participating states choose to go forward with management of wild American ginseng, land managers and owners of forestland can significantly affect the fate of wild ginseng through the tested procedure called stewarding. Stewarding involves active intervention in the demographic processes of a natural population to benefit its long-term growth and resilience to stressors, especially harvest.

How to Steward Your Ginseng Population

The steps are as follows:

1. Find, then refind the population. Record the GPS location of a landmark in or near the population. Many smartphones now have GPS capability. Note, however, that GPS is less accurate under heavy tree canopies and in hilly terrain. A second approach is to create a phototrail, a notebook containing sequential photos accompanied by bearings and distances leading to the population from some known starting point.

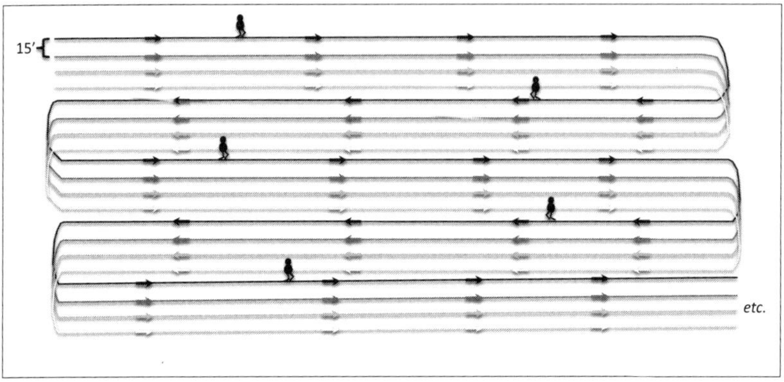

FIGURE 18.1.
Method of performing systematic sweeps across the forest to locate plants of an existing ginseng population. This particular sweep would use four people to walk the parallel transects, fifteen feet apart, marking all discovered plants with flagging. Orange pin flags, available at hardware stores, are ideal.

2. Once ginseng is found, perform broad sweeps through the population in spring to delineate the population (postgermination, ca. May 15–June 15). The approach involves walking in the center of fifteen-foot-wide parallel transects through the woods, temporarily marking ginseng plants with flagging (fig. 18.1). This procedure locates clusters of plants comprising a population.
3. Ginseng plants typically occur in clusters, groupings of from two to one hundred plants that occur within an area of a few square yards. Create a phototrail from each cluster to the next until all clusters are included, naming the clusters along the way (e.g., A, B, C, D, etc.).
4. Produce rough maps of each cluster to aid in refinding plants in the fall (see fig. 18.2). These maps do not need to be precise or carefully drawn but should include elements that will aid in refinding plants at a later date. Revisiting clusters in the fall and noting plants that have disappeared (e.g., due to browsing) can give an idea of the health of the population.
5. Gather basic data by cluster, including number of individuals in each leaf class to provide a baseline from which to judge effectiveness of stewarding in future years. (If time is an issue, steps 1–5 could be executed in the fall, but a more complete census is possible in the spring before tops have been lost to various midseason stresses.)
6. Plant seeds. The ideal seed planting time is after the majority of fruits are ripe (bright red) but before harvest season begins. In many states this will

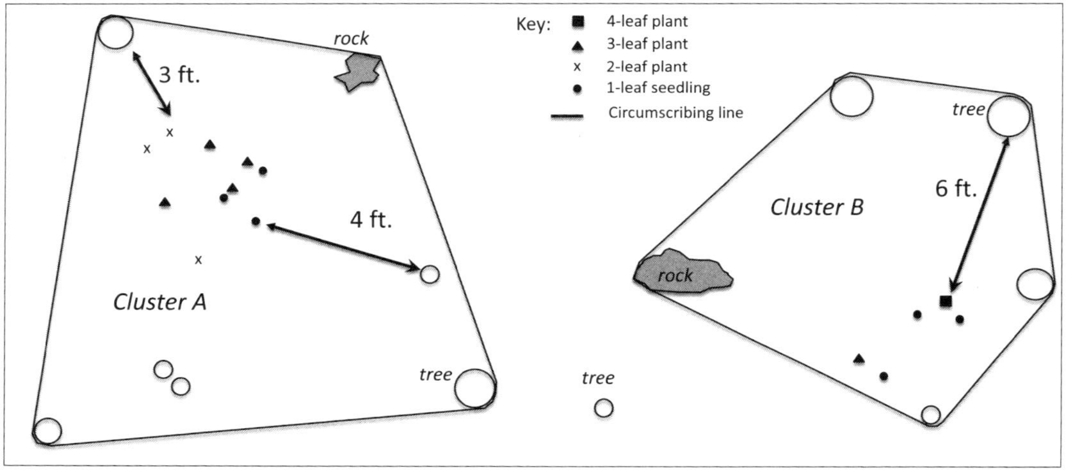

FIGURE 18.2.
Sample sketch map of ginseng clusters showing positions of individual plants relative to fixed objects such as large trees or rocks. Note different designations for ginseng plants of different size. Maps should be dated. New plants can be added over time. Adding approximate distances from objects can help in future relocation.

be late August, though population variation in fruit ripening is common. If most berries are still green, seed planting will need to be delayed.

- **a.** Seed depth. The optimal planting depth is one-half to one inch deep, depending on how loose the soil is. For firm soil, use one-half inch. For loose soil, one inch.
- **b.** Plant berries? Or seeds removed from berries? A proper experiment has not been performed to test which is optimal, but both will work. I suggest that if the berry has one seed, plant the whole berry. If the berry is two-seeded or three-seeded, remove the pulp and plant the seeds separately since only one seedling will likely succeed in a given location.
- **c.** Location considerations.
  - **(1)** Plant seeds one to three feet away from the parent plant to ensure that diseases or pests associated with the parent do not impact the seedling.
  - **(2)** Avoid planting seeds near decaying logs, where they may be more prone to fungal infection.
  - **(3)** Consider locations that naturally screen emerging plants from the reach of deer and the view of poachers.
  - **(4)** Heavily shaded micro-sites such as under spicebush may be suboptimal.

7. Clip tops. After seeds have been planted, carefully clip the stalk of large two-prong plants plus all three- or four-prong plants below the leaves. I recommend scissors for this task, not simply tugging on the stalk. Pulling on the stalk can break the underground rhizome and damage the plant. Since photosynthesis and restocking the root with carbohydrates is virtually complete by the end of August, clipping will not harm the plant. However, this is effective at hiding the most valuable adult plants from possible harvesters. Stalks should be discarded far from the population to avoid detection. (Some medicinal uses have been found for fresh leaf material.) Repeat steps six and seven annually to increase the natural population over time. Repeating the spring census after a few years can help gauge the success of stewarding.

Trial stewarding project. A trial stewarding project was begun in fall 2005 on a woodlot north of Morgantown, West Virginia, with a small natural population of ginseng (ca. 20–30 plants). In spring 2013 the population was thoroughly censused, and it had grown to 161 plants. In spring 2019 the population numbered 307 plants. Though ginseng is a slow-growing, low-seed-producing species, when seeds are carefully planted and harvest is discouraged, numbers can grow rapidly.

Although I have no plans to harvest in this population, by continuing this procedure I believe an offtake of 20–30 adults could be managed each year, focusing on older individuals whose legacy could be assured by planting seeds. If I were to harvest this population, I would also leave most seed-bearing individuals to maintain regeneration potential for the following year.

I would characterize the site where this stewarding project occurred as below average. The canopy showed signs of past disturbance, and soils were not well developed. While seeds germinate well, growth is slow, and none of the plants get very large. Deer browse and drought have been significant stressors in some summers.

If hundreds of landowners or land managers adopted population stewarding, the combination of these individual efforts could make a large difference over time. If all harvesters adopted stewarding as a way to make the practice sustainable, most likely harvest regulations of ginseng would be unnecessary.

Timber management. In addition to stewarding local natural populations, landowners and managers can take what we know about ginseng's response to the forest canopy to make better management decisions that benefit understory biodiversity. For example, if a woodlot is naturally maturing to over one hundred years of age, allowing the larger, older trees to naturally fall out and create gaps could stimulate native understory growth, ginseng in particular. If greater numbers of enlightened landowners viewed such forests as *increasing* in overall value, rather than as "overmature" (from a strictly forestry dollars-and-cents point of view), species such as ginseng will benefit.

If timber cutting is allowed, favor established forestry techniques that maintain a forest with diverse ages and sizes of trees (e.g., using single tree selection cutting or diameter-limit cutting with a high size threshold). These approaches create small gaps to allow light into the understory, while not being so large as to encourage overgrowth by more weedy elements such as blackberries, greenbrier, or, worse, invasive species. Careful attention should be made to minimizing soil disturbance caused by heavy machinery.

Deer. Deer herd management is also key to creating viable forest understory habitat. Encouraging responsible hunters on private land and prioritizing doe hunting can make a difference. In the vicinity of residences in the woods, encouraging safe bowhunting from fixed stands can be effective. Lacking sufficient hunting, landowners can support ecosystem balance by supporting reintroduction of top predators. Black bears and coyotes seem to be spreading on their own in much of the eastern forest. Even more effective would be reintroduction of wolves and mountain lions, since deer are more of a mainstay for their diets. If residents of Idaho, Montana, Wyoming, Colorado, California, Nevada, and Utah can coexist alongside top predators, shouldn't it be possible in the East as well? Public support will be key for making this happen.

Harvest. Finally, most states require harvesters to obtain permission from landowners to dig ginseng on their land. If you give permission, require harvesters to engage in stewardship harvesting practices. This includes harvesting only in season, only three-leaf or larger plants (preferably those with five leaflets on each leaf), and only those with red berries, planting the berries optimally on-site, and leaving any non-berry-producing adults plus at least a few large berry-producing plants to regenerate the population. I also suggest you require them to show you what they harvested, both as a check on their procedures but also to give you a baseline to judge whether future harvests are declining or not. If they show you pencil-thin roots, do not allow them to harvest again. If you do not allow harvesting on your land, report any suspicious activities or trespassing to local authorities to investigate. Harvesting without permission on your property is theft, and it should be treated as such. Finally, if you have ginseng on your land, do not tell anyone!

If you are not a landowner, I suggest making a financial contribution to a medicinal plant conservation organization such as United Plant Savers to help advance the cause of preserving valuable plants in their natural habitats.

Message to Land Managers

As protectors of our public and private natural heritage, land managers' efforts are key to conserving ginseng and other species in the diverse understory of eastern deciduous forests. Your lands often contain the most mature forests in a region, and many are not managed primarily for timber production and are therefore likely to become even more valuable to understory plants over time. Nevertheless, the strategy for managing public or private preserves needs to shift in the direction of holistic ecosystem management. Sometimes this is constrained by state laws. For example, deer hunting is prohibited in many state parks, resulting in severe overbrowsing. Recognizing this, some parks have obtained special permission to allow limited "controlled hunts," but often these are only taking place in the most extreme cases, where chronic overbrowsing has already desertified the understory. Managers can be advocates for earlier intervention to prevent lands from reaching this sad state. But nature is resilient; the understory will return, albeit slowly, if balance among plants, herbivores, and predators in the ecosystem is restored.

Public lands have been shown to be particularly vulnerable to ginseng poaching, which should be viewed as theft from the public treasury. Enforcement of no-harvest regulations is therefore key, and the temptation to overlook such assaults on our natural resources as less important than other human crimes should be avoided. When ginseng thieves are caught, confiscation of roots and restoration of populations is possible and is encouraged.

Site selection for replanting. Site selection for replanting poached ginseng involves a trade-off between ecological suitability and likelihood of further poaching. Given that ginseng, like most plants, shows local adaptation to the environment, replanting may be most successful in the site from which they were taken or a site with similar habitat characteristics (soils, overstory, aspect, elevation, and slope).

On the other hand, likelihood of being poached should be an additional criterion for site selection. First, if plants are returned to the site of poaching, planting there ensures that at least one poacher knows of its existence, and he or she may return. The availability of relatively inexpensive GPS units may increase that chance, even for remote populations. Second, if plants are put in an optimal site (but not the original site), a knowledgeable poacher will recognize and may target that site, though planting in poor sites is definitely not recommended.

Likewise, it may be possible to optimize the accessibility of replanted groups of plants relative to trails and roads. Plants too close to frequent human traffic will likely be discovered (and eventually poached). However, poachers want to elude discovery as well and therefore may stay away from trails and roads where their activities might be observed. The ideal location for replanted plants may be (1) far enough away from high-traffic areas that the plant will not be easily seen but (2) close enough that a human poacher might feel vulnerable to being caught. This seems like a hypothesis that needs to be tested; it is presently pure conjecture.

Replanting procedure. After locating suitable habitat, I recommend that roots be replanted in groups of ten to twenty individuals, with each plant separated from the neighboring plant by approximately twenty-five centimeters (ten inches). The spacing between individuals will reduce pathogen transmission among individuals. The spacing among groups should be approximately thirty to fifty feet, to make discovery less likely for someone walking through the woods yet keep groups in roughly the same forest community type. In addition, having variable locations for replanted groups will provide some insurance that at least some of them are in appropriate microenvironments.

Replanting the roots is simple:

1. A volume of native soil sufficient to accommodate the root should be dug with a trowel or small shovel, with a bit of extra depth if possible. Large rocks and roots should be removed where possible.
2. Holding the root nearly vertical with the rhizome at the top, position the root so the tip of the rhizome will end up one to two centimeters deep (one-half to one inch).
3. Sift loose soil gently around the root, trying to ensure good contact between the root and soil such that no air spaces are trapped, especially around the fine roots.
4. Cover the rhizome with fine organic soil and do not pack down. The angle of the rhizome does not matter (vertical, angled, or horizontal, depending on its natural tendency) as long as it is not buried too deep.

If dry weather is forecast for multiple days, a light watering will increase the probability of transplant success. No nutrients need to be added, but light weeding of neighboring (non-ginseng) plants could assist transplant success if there is dense crowding. Note that in the course of transporting roots (either by the harvester or replanting crew), rhizomes or root pieces may break off. These can sometimes regenerate whole plants, so plant these as well.

Expected replanting success rate. Two carefully performed transplant studies set the upper bounds of expected replanting success. In one experiment we harvested approximately ten individuals from each of ten populations. Each root was measured in terms of diameter and length, the rhizome was aged, and leaves were measured. Then the root was replanted back into the soil from which it was dug. In this study, 83 percent of the plants were present aboveground one year later. This underestimates transplant survival because some unknown fraction of plants that were not present the following year were alive but had either been browsed or had the bud damaged prior to censusing.

In a second study, plants were dug and then sterilized and transported in coolers to use in a growth chamber study. The plants were out of the ground four to five days before being replanted into pots, where they were grown through two growing seasons. At the end of the experiment, 97 percent of the plants were still alive. Replanting confiscated roots would be expected to have lower rates of success than either of the two studies above because once they are extracted, roots can start to become desiccated and fine root tips may break off. A third study approximates conditions more similar to the replanting of confiscated roots. In 2012, Turner-Skoff and McGraw transplanted 1,222 ginseng roots into a large field experiment at two widely separate sites, each site having mid-successional second-growth forest and mature late successional forest types. Roots had been dug from commercial cultivated herb gardens and shipped over a several-day period, then stored under refrigeration. Although the roots were sterilized, they had been out of the ground for nearly two weeks; though living, they were stressed by the transplant process in a manner similar to what harvested roots might experience. Roots were carefully planted in the field and followed over four growing seasons. Persistence rates after four years suggested survival rates of 72 to 79 percent annually.

Monitoring and stewarding restored populations. Use the methods described above for recording and stewarding populations back to healthy population sizes.

Message to Harvesters

While harvest is only one reason for decline of wild ginseng populations, it is one of the top-three causes, and the research has shown that harvester behaviors become even more critical in the face of the other stresses ginseng faces. If you are truly concerned about the long-term welfare of ginseng (and every harvester should be), I offer the following suggestions, in order of effectiveness.

1. **Grow it.** Consider becoming a forest grower and harvester of wild-simulated or woods-cultivated ginseng. If you choose this path, follow the guidance given to growers below to obtain roots that resemble wild and obtain the highest possible price. The payback time is longer for this, but by doing this you are providing a sustainable substitute for wild harvest and taking pressure off natural populations. If you are not a landowner, consider finding a family member or friend who is, who would allow you to grow ginseng on their property, possibly with a cost-share and profit-share for your operation.
2. **Steward it.** If you know of wild populations, consider stewarding them prior to harvesting. Assess the initial population size, grow the population to a target size using stewarding techniques described above, and harvest sustainably, always keeping the remaining population size above the initial size.

3. **Harvest ethically.** First, never harvest where harvest is explicitly prohibited. For private property, always obtain written permission of the landowner and carry it with you. Comply with season requirements of your state; seasons are set to ensure the opportunity for harvesters to plant viable seeds, reducing the effect of adult plant removal. Dig only three-prong or larger plants.

   Studies have shown that simply meeting these legal requirements is not enough for a sustainable harvest. The following steps will ensure that you are an ethical harvester who is not harming the species:

   a. Only harvest adult plants with red berries.
   b. Plant seeds optimally (one-half inch to one inch deep; one to three feet from parent plant; one seed per location).
   c. Plant all seeds you see, but leave a few of the largest plants unharvested; this ensures the future reproductive capacity of the population.
   d. Remove tops of unharvested adult plants to hide them from other harvesters.
   e. Establish a harvest cycle for each population (e.g., once every five years); consider stewarding the population in nonharvest years (planting seeds, clipping tops).
   f. At the time of this writing, the age minimum for harvest is five years, but this is insufficient time for plants to produce berries and seeds and replace themselves. Consider replanting any roots that are less than twelve years old, even if they technically meet the size requirements for harvest.
   g. Keep up to date on changes in harvest regulations.

4. **Educate others.** Encouraging ethical behavior in your fellow harvesters can help retain and restore the resource for all.
5. **If harvest becomes illegal.** Barring a dramatic change in policy (see chapter 15), eventually the species will be classified as rare everywhere if ginseng population trends continue as they have, and it will get the dubious honor of Endangered Species Act listing or listing in CITES Appendix I, demanding the highest level of protection. Harvest would need to cease. What then for a harvester? Become a grower. Most importantly, avoid the temptation to continue wild harvest. A black market may well develop, but penalties will no doubt be much more severe for wild harvest if this activity is banned.

Message to Growers

Forest growers are potentially the heroes of our story. As long as demand for wild-looking roots persists, ginseng growers who employ wild-simulated methods could be the salvation of truly wild populations by satisfying product demand. The first key to these efforts will be using local or regional seed sources wherever possible. This not only increases the probability of reaping a profit from the investment, it helps conserve wild populations by not spreading genes selected under cultivation and therefore bearing the wrong traits for your region. The second key will be using knowledge about the growth habits of wild ginseng described in this book to produce roots mimicking wild ones; this includes minimizing use of soil amendments, planting at low densities in minimally disturbed soils, and avoiding use of pesticides and fungicides. The third key will be emphasizing the sustainable, organic practices that add value to the final product, marketing forest-grown techniques, while making clear the conservation value-added basis of your growing methods. You may be able to add further value by demonstrating quality control with biochemical certification of the ginsenoside concentrations of your roots, as well as taking advantage of regional branding techniques (e.g., "Kentucky forest-grown ginseng," like "Vermont maple syrup").

I recommend that growers become well-informed about previous efforts to do forest-grown ginseng. The premier book on this topic is *Growing and Marketing Ginseng, Goldenseal and Other Woodland Medicinals*, by Jeanine Davis and W. Scott Persons. In addition, attending workshops with other growers can be hugely beneficial. Watch for these, led by organizations such as United Plant Savers (unitedplantsavers.org). United Plant Savers has a Forest Grown Verification Program that provides third-party verification "for non-timber forest grown products that are produced and harvested in a sustainable and legal manner."

Message for Asian Consumers

In her book, *Mending the Web of Life*, Elizabeth Call alludes to the proud two-thousand-year-plus history of Chinese medicine and emphasizes the need for practitioners to take an active role in conservation. This role has taken on special urgency as the human population has grown, while the sources of natural remedies have dwindled.

Supply. Demand. It is basic economics. Without demand from the Asian traditional medicine market, no supply is needed. The demand for wild root can be directly traced to an intangible and questionable value added to roots that are gnarly, wrinkled, or human shaped. While small differences in ginsenoside composition and concentrations can be found among ginseng roots on the basis of growth method, there is no established scientific basis for a difference in health effects of wild and cultivated ginseng.

To be blunt, the traditional Chinese medicine (TCM) market is responsible for threats to numerous wild species of animals and plants. Animals such as tigers, rhinoceroses, pangolins, tokay geckos, bears, and sea horses are killed to satisfy demand in numbers sufficient to threaten species as a whole. The taking of some of these animals is believed to be the origins of global pandemics (e.g., COVID-19). While the wild *plant* trade may not cause pandemics, long-term market demand drove Asian *Panax ginseng* to virtual extinction in the wild, and it is contributing to the same fate for American ginseng. As a scientist, while I respect the connection to nature and holistic values of TCM practitioners, I urge them to consider what the biochemistry says—that forest-grown and even cultivated ginseng can substitute very effectively for the wild harvest. As wise spenders, you should refuse to pay more at the point of sale for "wild-harvested" ginseng roots.

If wild ginseng harvest is eventually banned, refuse to purchase black market wild American ginseng.

Message for Scientists

There is much left to learn about *Panax quinquefolius* L. Here is a brief and by no means exhaustive list of the low-hanging fruit for future research.

Biochemical. While there have been many studies of ginsenoside composition in ginseng, little is known about controls over composition and concentrations as a function of genotype, ecotype, age, cultivation method, or growth environment. Development of inexpensive methods for characterization would allow growers to provide value-added information about their crop.

Medicinal. Methods used in evaluating efficacy of western medicines are challenging to apply to traditional Chinese medicine. Nevertheless, such attempts could be valuable in making the transition to more widespread acceptance or rejection of using traditional medicines such as ginseng. At the same time, rigorous evaluation, with proper controls, of specific ginsenosides to treat specific conditions could lead to breakthroughs for several important conditions, with FDA sanctioning.

- Can ginsenosides individually or in combination effectively treat or cure breast or prostate cancer, diabetes, brain injury, depression, neural degeneration, allergic reactions, or inflammation?
- Can ginsenosides boost immune response?
- Can ginsenosides suppress appetite and aid in weight loss?
- Can ginsenosides increase physical stamina?

Publications exist on all of these topics, yet western medicine has been slow to adopt any of these remedies.

Physiological and agricultural ecology

- Little is known about mutualistic (e.g., mycorrhizal) and pathogenic (especially fungal) relationships of ginseng.
- While it is clear that ginseng grows in a wide range of soils, rigorous experimental studies of soil effects on ginseng growth, survival and reproduction are needed.
- Is there evidence that loss of cations from soils caused by acid rain has negatively affected ginseng?
- Do ozone or other air pollutants affect ginseng?
- Why does growing ginseng at high density for several years leave soil in a condition unsuitable for further propagation after harvest?
- What is the best approach to developing local seed sources for those wanting to produce forest-grown ginseng?
- Can germination time be shortened?
- If special genotypes are found with desirable properties, can cloning methods be used to facilitate their propagation?
- Activated carbon appears to stimulate growth of ginseng in prior experiments. Why? Are there organic chemical compounds or other pollutants in the soil that are adsorbed by activated carbon?
- How do natural levels of fungal pathogens affect wild ginseng at normal densities? Is there evidence that fallen woody debris or katabatic winds affect pathogen load?

Population biology. Ginseng is one of the best-studied species in the world with respect to survival, growth, and reproduction of individual plants. With the long-term data sets collected by the McGraw lab from 1998 to 2016 available for scientific users, our hope is these data could be exploited by others over and over, using new and better methods as they are developed. Field census data described in this book are difficult to acquire for any species and unlikely to be repeated any time soon. Therefore, exploiting it to the fullest to judge population growth and viability is welcomed by all involved in the original effort.

- Using our knowledge of gravity-dispersal and thrush-dispersal, can a spatially explicit model of ginseng populations be developed to test the influences of these dispersal types on pattern development and test the importance of bird dispersal for cluster patterns?
- Why is survival rate of ginseng seedlings so high?
- Why does survival drop off with ages above around twenty-five years?
- What is the frequency of "true dormancy"—ginseng roots remaining alive belowground and *not* forming an aboveground top—and does this vary with environment or genotype? (Answering these questions would require earlier censusing than we performed each spring).
- We already know that resources control size and reproductive potential, but to what extent do resources (water, nutrients, and light) and/or pollinators control seed production rates (fraction of flower buds forming seeds)?
- Do seedlings risk death by germinating in close proximity to parent plants? If so, what is the cause?

Community and ecosystem ecology

- For forest-grown ginseng, strategies for growing in mixed stands (with other herbaceous species mixed in) need to be compared to strategies for growing in pure stands (just ginseng) with rigorous experiments.
- Which invasive plants have negative competitive effects on ginseng, and by what mechanisms do these effects work?
- Do nonnative earthworms affect ginseng success?
- Are there other common bird dispersers of ginseng seeds besides members of the thrush family?
- What role do mammalian omnivores such as raccoons or bears play in infrequent long-distance seed dispersal?
- Does dispersal in dung alter seedling success rates?
- Do thrushes use ginseng as an energy and stamina source, much like humans do?
- Do any other animals self-medicate with ginsenosides? Deer? Bears? Chipmunks?

Genetics

- The genome of *Panax quinquefolius* needs to be sequenced to aid in many kinds of genetic studies.
- What are the key differences that have evolved under cultivation?

- Do these differences result in outbreeding depression if crosses occur with wild populations?
- How severe is inbreeding depression (beyond initial seedling phases)?
- Can reliable markers be developed for cultivated ginseng to detect presence in wild populations?
- Are locally adapted genotypes already displaced from their climatic optimum conditions?
- Could assisted migration benefit ginseng conservation in future climates?
- Is there heritable variation for size or growth in natural populations?
- Is there a trade-off between growth rate and production of secondary compounds such as ginsenosides?

ACKNOWLEDGMENTS

First and foremost, I thank my wife, Ann, for her support during my whole career but especially during the twenty years from 1998 to 2018 when I left home each summer to lead field research crews with my graduate and undergraduate students. This meant I could ensure the continuity of methods and training necessary to maintain the quality of the research for such a long-term study. This work would not have been possible without her. My children also tolerated my absence, and my daughter Caitlin, a biology major, stepped in on multiple occasions to help with field censusing. Thank you.

This research would not have been possible without the introduction to wild-harvested plants I received from Martha Van der Voort and Brent Bailey in the mid-1990s. Martha's subsequent PhD work on harvest launched our demographic studies on ginseng. PhD students Mary Ann Furedi and Emily Mooney provided the key early legwork to locate our long-term study populations and work out the kinks in the censusing protocols. PhD students Sara Souther, Jessica Turner-Skoff, and Jennifer Chandler in turn assumed the mantle of Conservation Fellow and managed our lab website, developed training materials for undergraduates, led research teams, and performed significant public outreach. MS students Erin Hackney, Kerry Wixted, Zach Bradford, and Amy Hruska co-led research teams each summer. All of these graduate students were at the core of the success of our ginseng research program, and their work formed the bulk of the material presented in this book. I know I was a demanding advisor at times, but I had high expectations for all of them because I believed deeply in the individual potential of each. I thank them all from the bottom of my heart for their selfless dedication to the cause, their cheerful manner under trying conditions, and their commitment to quality research for the larger goal of understanding stressors in natural ecosystems. I also thank them for letting me introduce them individually to a broader audience. I did not exaggerate in the text—these are all wonderful human beings, heroes for science and the planet.

Colleague and friend Dr. Anne Lubbers from Centre College, Kentucky, led census teams from the early days of our fieldwork, and, as importantly, graciously hosted our WVU teams at her home in Danville. After grueling days of fieldwork and camping, her hospitality, real mattresses, and mint juleps were a godsend. My former grad student Dr. Rob Lamar did the same on a few of our Virginia loops.

At various times, but especially at meetings centered on ginseng, I have benefited from interactions with my ginseng-centric colleagues around the country, including, but not limited to, Bob Beyfuss, Paul Hsu, Susan Leopold, Daniel Gagnon, Andree Nault, Eric Burkhart, Roger Anderson, Dan Drees, Susan Farrington, Jennifer Cruse-Sanders, Scott Persons, John Young, Janet Rock, and Jeanine Davis.

Undergraduate conservation interns did a lot of the difficult work of field research alongside graduate students and me. They all worked hard in the field and lab and tolerated my attempts to teach them bird songs and an appreciation for prog rock. Several of them developed fine independent research projects (honors theses) on ginseng. Though I fear this list is not complete, they include April Stark, Erica Dakan, Harmony Burwell, Mindi Spencer, Kristen Hunter-Cevera, Chris Packert, Mary Olive, Britni Schoonover, Stephanie Hovatter, Nathaniel Lee, Adam Martin, Rebecca Kenyon, Clare Maloy, Stevia Morawski, Mark Guido, Allison Kenyon, Lorraine Carter-Lovejoy, Diana Black, Alix Wagner, John Hall, Zac Zacavish, Michael Elza, Kyle Gregory, Anne Jarrell, Quinn Doyle, Mark Arbuiso, Jake Richards, Chris Mallow, and Gera Jochum. Emily Thyroff played a key role in curating and error-correcting the data set.

Two anonymous manuscript reviewers made valuable suggestions to improve the manuscript.

Components of the research described in this book were funded by the U.S. Geological Survey, the U.S. Fish & Wildlife Service, and Phipps Conservatory. The bulk of the funding came from the National Science Foundation. In particular, the Long-Term Research in Environmental Biology program at NSF was critical for allowing the ongoing censusing efforts that led to so many of the findings discussed here. Most of the conclusions we reached depended on the long-term nature of the program. I never forgot that NSF is funded by taxpayers. The years spent compiling this book were, in part, my effort to give back to those who ultimately funded the funders of this research. You deserve to know about, and benefit from, your investment in science. The data set underpinning much of this research has been published and is available to the public as well: J. B. McGraw, M. E. Van der Voort, M. A. Furedi, A. E. Lubbers, E. Mooney, S. Souther, J. Turner-Skoff, J. L. Chandler, and E. Thyroff, "Long Term Research in Environmental Biology: Demographic Census Data for Thirty Natural Populations of American Ginseng, 1998–2016," Long Term Ecological Research Network, 2017.

I am also indebted to the many private landowners, nature preserve managers, and state and national land managers who generously allowed us access to their properties every year to continue this effort. Locations and specific sites are purposely obscured to protect the ginseng populations.

BIBLIOGRAPHY

Chapter 1. The Panacea

Bernards, M. A., L. F. Yousef, and R. W. Nicol. 2006. The allelopathic potential of ginsenosides. In: Inderjit and K. G. Mukerji, editors. *Allelochemicals: Biological Control of Plant Pathogens and Diseases*, Disease Management of Fruits and Vegetables 2. Dordrecht: Springer. Pp. 157–175.

Christensen, Lars P. 2008. Ginsenosides: Chemistry, biosynthesis, analysis, and potential health effects. *Advances in Food and Nutrition Research* 55: 1–99.

Dharmananda, Subhuti. 2002. The nature of ginseng: Traditional use, modern research, and the question of dosage. *HerbalGram* 54: 34–51.

Xiang, Y. Z., H. C. Shang, X. M. Gao, and B. L. Zhang. 2008. A comparison of the ancient use of ginseng in traditional Chinese medicine with modern pharmacological experiments and clinical trials. *Phytotherapy Research* 22 (7): 851–858.

Chapter 2. The Romantic Ideal

Carlson, A. W. 1986. Ginseng: America's botanical drug connection to the Orient. *Economic Botany* 40 (2): 233–249.

Case, M. A., K. M. Flinn, J. Jancaitis, A. Alley, and A. Paxton. 2007. Declining abundance of American ginseng (*Panax quinquefolius* L.) documented by herbarium specimens. *Biological Conservation* 134: 22–30.

Hufford, M. 1999. Tending the commons: Folklife and landscape in southern West Virginia. Library of Congress. https://www.loc.gov/collections/folklife-and-landscape-in-southern-west-virginia/about-this-collection.

Johannsen, K. 2006. *Ginseng Dreams: The Secret World of America's Most Valuable Plant*. Lexington: University Press of Kentucky.

Manget, L. 2012. Sangin' in the mountains: The ginseng economy of the Southern Appalachians, 1865–1900. *Appalachian Journal* 40 (1–2): 28–56.

Manget, L. 2016. Nature's emporium: The botanical drug trade and the commons tradition in southern Appalachia, 1847–1917. *Environmental History* 21 (4): 660–687.

McGraw, J. B. 2001. Evidence for decline in stature of American ginseng plants from herbarium specimens. *Biological Conservation* 98: 25–32.

McGraw, J. B., S. M. Sanders, and M. Van der Voort. 2003. Distribution and abundance of *Hydrastis canadensis* L. (Ranunculuaceae) and *Panax quinquefolius* L. (Araliaceae) in the central Appalachian region. *Journal of the Torrey Botanical Society* 130 (2): 62–69.

Muir, J. 1916. A Thousand Mile Walk to the Gulf. Boston: Houghton Mifflin.

Taylor, D. A. 2006. *Ginseng: The Divine Root.* Chapel Hill: Algonquin Books of Chapel Hill.

Wang, D. 2012. American ginseng and its effect on Americanization. *Virginia Review of Asian Studies* 1: 131–135.

Chapter 3. John Harper's Dream Plant

Grant, M. C. 1993. The trembling giant. *Discover* 14 (10): 82–89.

Harper, J. L. 1977. *Population Biology of Plants.* London: Academic Press.

McGraw, J. B., A. E. Lubbers, M. E. Van der Voort, E. H. Mooney, M. A. Furedi, S. Souther, J. B. Turner, and J. Chandler. 2013. Ecology and conservation of ginseng (*Panax quinquefolius*) in a changing world. *Annals of the New York Academy of Sciences* 1286: 62–91.

Mooney, E. H., and J. B. McGraw. 2007. Effects of self-pollination and outcrossing with cultivated plants in small natural populations of American ginseng, *Panax quinquefolius* (Araliaceae). *American Journal of Botany* 94 (10): 1677–1687.

Chapter 4. A Species in Decline?

Hutchings, M. J. 1991. Monitoring plant populations: Census as an aid to conservation. In: F. B. Goldsmith, ed., *Monitoring for Conservation and Ecology.* London: Chapman and Hall. Pp. 61–76.

Chapter 5. From Monitoring to Demography

Caswell, H. 2001. *Matrix Population Models.* 2nd ed. Sunderland, Mass.: Sinauer.

Charron, D., and D. Gagnon. 1991. The demography of northern populations of *Panax quinquefolium* (American ginseng). *Journal of Ecology* 79 (2): 431–445.

McGraw, J. B., M. E. Van der Voort, Mary Ann Furedi, A. E. Lubbers, E. Mooney, S. Souther, J. Turner-Skoff, J. L. Chandler, and E. Thyroff. 2017. Long term research in environmental biology: Demographic census data for thirty natural populations of American Ginseng: 1998–2016. Long Term Ecological Research Network.

Chapter 6. The Root of the Harvest Problem

Bailey, B. 1999. Social and economic impacts of wild harvested products. PhD diss., West Virginia University.

McGraw, J. B., et al. 2005. Berry ripening and harvest season in wild American ginseng. *Northeastern Naturalist* 12 (2): 141–152.

Nantel, P., D. Gagnon, and A. Nault. 1996. Population viability analysis of American ginseng and wild leek harvested in stochastic environments. *Conservation Biology* 10 (2): 608–621.

Schmidt, J. P., J. Cruse-Sanders, J. L. Chamberlain, S. Ferreira, and J. A. Young. 2019. Explaining harvests of wild-harvested herbaceous plants: American ginseng as a case study. *Biological Conservation* 231: 139–149.

Van der Voort, M. E. 1998. An inventory of wild-harvested plants in the Otter Creek Wilderness Area of the Monongahela National Forest, West Virginia. Master's thesis, West Virginia University.

Van der Voort, M. E. 2005. An ecological study of *Panax quinquefolius* in central Appalachia: Seedling growth, harvest impacts and geographic variation in demography. PhD diss., West Virginia University.
Van der Voort, M. E., and J. B. McGraw. 2006. Effects of harvester behavior on population growth rate affects sustainability of ginseng trade. *Biological Conservation* 130: 505–516.

Chapter 7. The Case of the Disappearing Plants

Farrington, S. J., R-M. Muzika, D. Drees, and T. M. Knight. 2009. Interactive effects of harvest and deer herbivory on the population dynamics of American ginseng. *Conservation Biology* 23 (3): 719–728.
Furedi, M. A. 2004. Effects of herbivory by white-tailed deer (*Odocoileus virginianus* Zimm.) on the population ecology and conservation biology of American ginseng (*Panax quinquefolius* L.). PhD. diss., West Virginia University.
Furedi, M. A., and J. B. McGraw. 2004. White-tailed deer: Dispersers or predators of American ginseng seeds? *American Midland Naturalist* 152: 268–276.
McGraw, J. B., and J. L. Chandler. 2018. Demographic hallmarks of an overbrowsed population state in American ginseng. *Global Ecology and Conservation* 15: 1–12.
McGraw, J. B., and M. A. Furedi. 2005. Deer browsing and population viability of a forest understory plant. *Science* 307 (5711): 920–922.

Chapter 8. The Elephant in the Room

Global Climate Change. National Aeronautics and Space Administration. Climate.nasa.gov/.
Souther, S. 2011. Demographic response of American ginseng (*Panax quinquefolius* L.) to climate change. PhD diss, West Virginia University.
Souther, S., M. J. Lechowicz, and J. B. McGraw. 2012. Experimental test for adaptive differentiation of ginseng populations reveals complex response to temperature. *Annals of Botany* 110: 829–837.
Souther, S., and J. B. McGraw. 2011. Local adaptation to temperature and its implications for species conservation in a changing climate. *Conservation Biology* 25 (5): 922–931.
Souther, S., and J. B. McGraw. 2011. Vulnerability of wild American ginseng to an extreme early spring temperature fluctuation. *Population Ecology* 53 (1): 119–129.
Souther, S., and J. B. McGraw. 2014. Synergistic effects of climate change and harvest on extinction risk of American ginseng. *Ecological Applications* 24 (6): 1463–1477.

Chapter 9. Appalachian Outlaws

Hardin, G. 1968. The tragedy of the commons. *Science* 162 (3859): 1243–1248.
McGraw, J. B., S. Souther, and A. E. Lubbers. 2010. Rates of harvest and compliance with regulations in natural populations of American ginseng (*Panax quinquefolius* L.). *Natural Areas Journal* 30: 202–210.
Van der Voort, M. E., B. Bailey, D. E. Samuel, and J. B. McGraw. 2003. Recovery of populations of goldenseal (*Hydrastis canadensis* L.) and American ginseng (*Panax quinquefolius* L.) following harvest. *American Midland Naturalist* 149: 282–292.

Chapter 10. The Ever-Changing Landscape

Abumrad, J. 2020. Dolly Parton's America. www.dollypartonsamerica.com.

Britton-Purdy, J. 2016. The violent remaking of Appalachia. *Atlantic*. https://www.theatlantic.com/technology/archive/2016/03/the-violent-remaking-of-appalachia/474603.

Chandler, J. L., and J. B. McGraw. 2015. Variable effects of timber harvest on the survival, growth, and reproduction of American ginseng (*Panax quinquefolius* L.). *Forest Ecology and Management* 344: 1–9.

Chandler, J. L. 2016. Ecological effects of forest canopy disturbance on the understory plant, American ginseng (*Panax quinquefolius* L.). PhD diss., West Virginia University.

Chandler, J. L. 2017. A slow opportunist: Physiological and growth responses of an obligate understory plant to patch cut harvesting. *Oecologia* 183: 677–687.

Chandler, J. L., and J. B. McGraw. 2017. Demographic stimulation of the obligate understorey herb, *Panax quinquefolius* L., in response to natural forest canopy disturbances. *Journal of Ecology* 105 (3): 736–749.

Clarkson, R. B. 1964. *Tumult on the Mountains: Lumbering in West Virginia 1770–1920*. Parsons, W.V.: McClain Printing.

Cox, W. 2016. America's most urban states. *Newgeography*. https://www.newgeography.com/content/005187-america-s-most-urban-states.

Drummond, M. A., and T. R. Loveland. 2010. Land-use pressure and a transition to forest-cover loss in the eastern United States. *Bioscience* 60 (4): 286–298.

Flinn, K. M., and M. Vellend. 2005. Recovery of forest plant communities in post-agricultural landscapes. *Frontiers in Ecology and the Environment* 3 (5): 243–250.

Leopold, A. 1966. *A Sand County Almanac, with Essays on Conservation from Round River*. New York: Ballantine Books.

Pericak, A. A., C. J. Thomas, D. A. Kroodsma, M. F. Wasson, M. R. V Ross, N. E. Clinton, D. J. Campagna, Y. Franklin, E. S. Bernhardt, and J. F. Amos. 2018. Mapping the yearly extent of surface coal mining in Central Appalachia using Landsat and Google Earth Engine. PLOS ONE. https://doi.org/10.1371/journal.pone.0197758.

Steyaert, L. T., and R. G. Knox. 2008. Reconstructed historical land cover and biophysical parameters for studies of land-atmosphere interactions within the eastern United States. *Climate and Dynamics* 113 (D2): 1–27.

Turner, J. B. 2015. The root of sustainability: Investigating the relationship between medicinal plant conservation and surface mining in Appalachia. PhD diss., West Virginia University.

Wagner, A., and J. B. McGraw. 2013. Sunfleck effects on physiology, growth, and local demography of American ginseng (*Panax quinquefolius* L.). *Forest Ecology and Management* 291: 220–227.

Chapter 11. Space Invaders (and Other Pests)

Callaway, R. M., and W. M. Ridenour. 2004. Novel weapons: Invasive success and the evolution of increased competitive ability. *Frontiers in Ecology and the Environment* 2 (8): 436–443.

Diagne, C., B. Leroy, A-C Vaissiere, R. E. Gozlan, D. Roiz, I. Jaric, J-M Salles, C. J. A. Bradshaw, and F. Courchamp. 2021. High and rising economic costs of biological invasions worldwide. *Nature* 592: 571–576.

Dlugosch, K. M., F. A. Cang, B. S. Barker, K. Andonian, S. M. Swope, and L. H. Rieseberg. 2015. Evolution of invasiveness through increased resource use in a vacant niche. *Nature Plants* 1: 15066.

Liu, H., and P. Stiling. Testing the enemy release hypothesis: A review and meta-analysis. *Biological Invasions* 8: 1535–1545.

Wixted, K. 2009. A Panax-centric view of invasive species and a case study on the effects of garlic mustard (*Alliaria petiolata*). Master's thesis, West Virginia University.

Wixted, K., and J. B. McGraw. 2009. A Panax-centric view of invasive species. *Biological Invasions* 11 (4): 883–893.

Wixted, K., and J. B. McGraw. 2010. Competitive and allelopathic effects of garlic mustard (*Alliaria petiolata*) on American ginseng (*Panax quinquefolius*). *Plant Ecology* 208: 347–357.

Chapter 12. Darwinian Forces to the Rescue?

Cruse-Sanders, J. M., and J. L. Hamrick. 2004. Genetic diversity in harvested and protected populations of wild American Ginseng, *Panax quinquefolius* L. (Araliaceae). *American Journal of Botany* 91 (4): 540–548.

Cruse-Sanders, J. M., J. L. Hamrick, and J. A. Ahumada. 2005. Consequences of harvesting for genetic diversity in American Ginseng (*Panax quinquefolius* L.): A simulation study. *Biodiversity and Conservation* 14: 493–504.

Darwin, C. 1859. *On the Origin of Species by Means of Natural Selection*. London: John Murray.

Mooney, E. H., and J. B. McGraw. 2007. Alteration of selection regime resulting from harvest of American ginseng, *Panax quinquefolius*. *Conservation Genetics* 8: 57–67.

Mooney, E. H., and J. B. McGraw. 2009. Relationship between age, size and reproduction in populations of American ginseng, *Panax quinquefolius* (Araliaceae), across a range of harvest pressures. *Ecoscience* 16 (1): 84–94.

Chapter 13. Dispersal to the Rescue?

Elza, M. C., C. Slover, and J. B. McGraw. 2016. Analysis of wood thrush (*Hylocichla mustelina*) movement patterns to explain spatial structure of American ginseng (*Panax quinquefolius*) populations. *Ecological Research* 31: 195–201.

Hackney, E. 1999. The effects of small population size, breeding system, and gene flow on fruit and seed production in American ginseng (*Panax quinquefolius* L., Araliaceae). Master's thesis, West Virginia University.

Hackney, E., and J. B. McGraw. 2001. Experimental demonstration of an Allee effect in American ginseng. *Conservation Biology* 15 (1): 129–136.

Hruska, A. M. 2014. Animal dispersal of American ginseng (*Panax quinquefolius* L.). Master's thesis, West Virginia University.

Hruska, A. M., S. Souther, and J. B. McGraw. 2014. Songbird dispersal of American ginseng (*Panax quinquefolius* L.). *Ecoscience* 21 (1): 1–10.

Meyer, A. 2016. Smithsonian scientists solve puzzle of dramatic wood thrush decline. *Smithsonian Insider*. https://insider.si.edu/2016/01/loss-breeding-grounds-north-america-likely-cause-wood-thrush-decline.

Rosenberg, K. V., et al. 2019. Decline of the North American avifauna. *Science* 366 (6461): 120–124.

Webb, S. L. 1986. Potential role of passenger pigeons and other vertebrates in the rapid Holocene migrations of nut trees. *Quaternary Research* 26: 367–375.

Chapter 14. Conservation in the Anthropocene

Davis, J., and W. Scott Persons. 2014. *Growing and Marketing Ginseng, Goldenseal and Other Woodland Medicinals.* Gabriola Island, British Columbia: New Society Publishers.

U.S. Fish and Wildlife Service. American ginseng. https://www.fws.gov/international/plants/american-ginseng.html.

Winston, M. L. 2014. *Bee Time: Lessons from the Hive.* Cambridge, Mass.: Harvard University Press.

Chapter 15. *Homo sapiens*; Wise Human?

Burkhart, E. P., and M. G. Jacobson. 2009. Transitioning from wild collection to forest cultivation of indigenous medicinal forest plants in eastern North America is constrained by lack of profitability. *Agroforestry Systems* 76: 437–453.

Leopold, S., and A. Ormsby. 2018. *Proceedings: Conservation, Cultivation, Commerce: The Future of Ginseng and Forest Botanicals Symposium, July 12–14, 2017, Morgantown, WV.* United Plant Savers.

INDEX